Deflection of Fracturing Fracture Network Disturbed by Discontinuity and Multiple Fractures in Rock

Yongliang Wang

Deflection of Fracturing Fracture Network Disturbed by Discontinuity and Multiple Fractures in Rock

Science Press
Beijing

Springer

Yongliang Wang
China University of Mining
and Technology (Beijing)
Beijing, China

ISBN 978-981-95-1568-4 ISBN 978-981-95-1569-1 (eBook)
https://doi.org/10.1007/978-981-95-1569-1

Jointly published with Science Press.
The print edition is not for sale in China mainland. Customers from China mainland please order the print book from: Science Press.
ISBN of the Co-Publisher's edition: 978-7-03-078560-2

This Springer imprint is published by the registered company Springer Nature Singapore Pte Ltd.
The registered company address is: 152 Beach Road, #21-01/04 Gateway East, Singapore 189721, Singapore

Preface

The deflection of fracturing fracture network disturbed by discontinuity and multiple fractures in rock is investigated in this book. Due to the complexity of fracturing problems in actual engineering, such as multiphysical fields coupling, multiscale fractures, reservoir rock discontinuity, and interactions between multiple fractures, numerical methods are more effective than theoretical and experimental methods; this book aims at the development of related numerical methods, models, and cases for conducting investigations on some scientific and engineering issues. It can provide a reference for those engaged in the research of propagation behaviours of fracturing fracture network, and have a comprehensive grasp of the research in this field.

The book consists of (1) dynamic intersection and deflection behaviours of hydraulic fractures meeting granules and natural fractures in tight reservoir rock based on statistical modelling and fractal characterization, (2) deflection behaviours and fractal morphology of hydraulic fractures meeting beddings and granules with variable geometrical configurations and geomechanical properties, (3) dynamic propagation of tensile and shear fractures induced by impact load in rock based on dual bilinear cohesive zone model, (4) centre- and edge-type intersections of hydraulic fracture network under varying crossed natural fractures and fluid injection rate, (5) wells connection and long hydraulic fracture induced by multi-well hydrofracturing utilizing crossed perforation clusters, (6) deflection of fracture networks and gas production in multi-well hydrofracturing utilizing parallel and crossed perforation clusters, (7) supercritical CO_2-driven intersections of multi-well fracturing fracture network and induced microseismic events in naturally fractured reservoir.

The author gratefully acknowledges the financial support from the research projects led by the author, i.e. the National Natural Science Foundation of China (Grant Nos. 41877275 and 51608301), the Beijing Natural Science Foundation (Grant L212016), the China National Petroleum Corporation (CNPC) Innovation Found (Grant No. 2022DQ02-0204), the Fundamental Research Funds for the Central Universities, Ministry of Education of China (Grant Nos. 2023JCCXLJ04 and 2019QL02), the Open Project Foundation of National Engineering Research Center of Oral Biomaterials and Digital Medical Devices (Grant No. PKUSS20240202), the China Postdoctoral Science Foundation (Grant Nos. 2018T110158, 2016M601170, and 2015M571030), the Key Laboratory Open Project Foundation of Soft Soil Characteristics and Engineering Environment (Grant No. 2017SCEEKL003), the Teaching Reform and Research Projects of Undergraduate Education, CUMTB (Grant Nos. J241505, J210613, J200709, and J190701), the Innovation Training Projects for Undergraduates, CUMTB (Grant Nos. 202415019, 202306002, 202306011, 202206005, 202106001, and 202106030), and the Yue Qi Young Scholar Project Foundation, CUMTB (Grant No. 2019QN14).

The author gratefully acknowledges the guidance and advice from the respectable tutors during the master, Ph.D., and postdoctorate stages, Prof. Yuan Si and Prof. Zhuang Zhuo of Tsinghua University, and Prof. Wu Jianxun and Prof. Ju Yang of the China University of Mining and Technology (Beijing). During the postdoctorate stage, the author visited several research centres for computational mechanics in famous foreign universities as visiting scholar; the author gratefully acknowledges the advice and comments from the collaborators, Prof. Li Chenfeng, Prof. Feng Yuntian, and Prof. D. Roger J. Owen of the Zienkiewicz Centre for Computational Engineering at Swansea University in the UK, Prof. David Kennedy and Prof. Frederic W. Williams of the Applied and Computational Mechanics Group at Cardiff University in the UK, John Cain, Melanie Armstrong, and Fen Paw at Rockfield Software Ltd. in the UK, and Prof. Robert L. Taylor of the Department of Civil and Environmental Engineering at the University of California, Berkeley, in the USA. The author gratefully acknowledges the participation and work from the graduate students in the research group, who contributed their intelligences. The author also gratefully acknowledges the editors at the Science Press in China, for providing many suggestions and much assistance on formatting modifications and typesetting adjustments for improving this manuscript.

Because this book is restricted by the limited knowledge of the author, a few errors are unavoidable. The author hopes all that experts, scholars, and other readers of this book will provide helpful suggestions for the book's improvement.

January 2024

Dr. Yongliang Wang
Head of Computational Mechanics Group
Department of Engineering Mechanics
School of Mechanics and Civil Engineering
State Key Laboratory for Fine Exploration
and Intelligent Development of Coal Resources
China University of Mining and Technology (Beijing)
Beijing, China
wangyl@cumtb.edu.cn
https://www.wangyongliang.net

Competing Interests The author has no competing interests to declare that are relevant to the content of this manuscript.

About This Book

The extraction of deep unconventional oil and gas resources is a crucial support for the future utilization of energy. The hydrofracturing, which forms a complex fracture network in deep reservoir rock, is a key technology for oil and gas flow and extraction. The propagation behaviours (such as deflection, penetration, and intersection) and fracture network morphology are crucial to the evaluation and control of oil and gas production. This book mainly introduces the following three aspects, including deflection of fracturing fracture network disturbed by discontinuity (beddings, granules, and natural fractures) in rock, deflection of fracturing fracture network disturbed by multiple fractures, propagation, and deflection of fracturing fracture network in supercritical CO_2 fracturing. The book focuses on the numerical research progresses of deflection of fracturing fracture network disturbed by discontinuity and multiple fractures in rock, which covers the following main contents: (1) dynamic intersection and deflection behaviours of hydraulic fractures meeting granules and natural fractures in tight reservoir rock based on statistical modelling and fractal characterization, (2) deflection behaviours and fractal morphology of hydraulic fractures meeting beddings and granules with variable geometrical configurations and geomechanical properties, (3) dynamic propagation of tensile and shear fractures induced by impact load in rock based on dual bilinear cohesive zone model, (4) centre- and edge-type intersections of hydraulic fracture network under varying crossed natural fractures and fluid injection rate, (5) wells connection and long hydraulic fracture induced by multi-well hydrofracturing utilizing crossed perforation clusters, (6) deflection of fracture networks and gas production in multi-well hydrofracturing utilizing parallel and crossed perforation clusters, and (7) supercritical CO_2-driven intersections of multi-well fracturing fracture network and induced microseismic events in naturally fractured reservoir.

Given its scope, the book offers a valuable reference guide for researchers, postgraduates, and undergraduates majoring in engineering mechanics, mining engineering, petroleum engineering, geotechnical engineering, and geological engineering.

Contents

About the Author

Dr. Yongliang Wang is currently a researcher in the Department of Engineering Mechanics, School of Mechanics and Civil Engineering, State Key Laboratory for Fine Exploration and Intelligent Development of Coal Resources, at China University of Mining and Technology (Beijing), and the head of Computational Mechanics Group. He obtained his Ph.D. degree from the Department of Civil Engineering at Tsinghua University in 2014. In 2015, 2016, 2017, and 2019, he successively visited the Zienkiewicz Centre for Computational Engineering at Swansea University, UK, the Applied and Computational Mechanics Center at Cardiff University, UK, and the Rockfield Software Ltd., UK, to carry out cooperative research. In 2022 and 2023, he visited the University of California, Berkeley and San Diego, USA, as a visiting scholar.

His research interests include high-performance adaptive finite element methods, computation and analysis of rock damage and fracture, and structural vibration and stability. He has also taught computational mechanics at the undergraduate level and the basic theory of the finite element method, computational solid mechanics, and rock fracture mechanics, and frontier and progress in mechanics at the graduate level. He is a Board Member of the Soft Rock Branch of the Chinese Society for Rock Mechanics and Engineering and a Member of the Chinese Society of Theoretical and Applied Mechanics, China Civil Engineering Society, and China Coal Society; moreover, he serves as a project expert of the National Natural Science Foundation of China, and project expert of the Degree Center of the Ministry of Education of China. He serves as a project expert of the National Natural Science Foundation of China, and project expert of the Degree Center of the Ministry of Education of China; moreover, he is the Editorial Board Member, Youth Editorial Board Mmember, and Guest Editor of the *Journal of Petroleum and Petrochemical Engineering*, *Journal of Intelligent Construction*, *Green and Smart Mining Engineering*, *Energy and Fuel*, *Energies*, *Frontiers in Build Environment*, *Journal of Engineering Science*, etc.

As the person in charge, he presided over 18 research projects, including the National Natural Science Foundation of China, Beijing Natural Science Foundation, China National Petroleum Corporation (CNPC) Innovation Foundation, Fundamental Research Funds for the Central Universities, Ministry of Education of China,

Open Project Foundation of National Engineering Research Center of Oral Biomaterials and Digital Medical Devices, China Postdoctoral Science Foundation, Key Laboratory Open Project Foundation of Soft Soil Characteristics and Engineering Environment, Teaching Reform and Research Projects of Undergraduate Education, and Yue Qi Young Scholar Project Foundation. He participated in 12 research projects, including the Major Scientific Research Instrument Development of the National Natural Science Foundation and the National Key Research and Development Program of China. He published 5 books in English and more than 190 academic papers; furthermore, he obtained more than 50 software copyrights. He has received the Science and Technology Invention Award, Chinese Society for Rock Mechanics and Engineering (2024), the Science and Technology Advancement Award, China Petroleum and Chemical Industry Federation (2024), the Science and Technology Advancement Award, Zhongguancun Green Mine Industry Alliance (2024), the Rock Mechanics Education Award, Chinese Society for Rock Mechanics and Engineering (2021), the high-quality book award for Beijing universities (2023), the Excellent Supervisor Award, CUMTB (2019, 2020), the Science and Technology Award, China Coal Industry Association (2019), the Yue Qi Young Scholar Award, CUMTB (2019), the Emerald Literate Highly Commended Paper Award (2018), and the Frontrunner 5000 (F5000) Top Articles in Outstanding S&T Journals of China (2016, 2022).

Chapter 1
Introduction

1.1 Introduction

The extraction of deep unconventional oil and gas resources (such as coal bed methane and shale gas) is a crucial support for the future utilization of energy. The stimulated reservoir volume (i.e., hydraulic fracturing or hydrofracturing), which forms a complex fracture network in deep reservoir rock, is a key technology for oil and gas flow and extraction. The propagation behaviors (such as deflection, penetration, and intersection) and fracture network morphology are crucial to the evaluation and control of oil and gas production; the novel supercritical CO_2 (SC-CO_2) fracturing technology has the potential to increase the complexity of fracture network, reduce pollution and improve oil and gas recovery, which is also discussed. This book mainly introduces the following three aspects, including: deflection of fracturing fracture network disturbed by discontinuity (beddings, granules, and natural fractures) in rock, deflection of fracturing fracture network disturbed by multiple fractures in rock, propagation and deflection of fracturing fracture network in SC-CO_2 fracturing. Due to the complexity of fracturing problems in actual engineering, such as multiphysical fields coupling, multiscale fractures, reservoir rock discontinuity, and interactions between multiple fractures, numerical methods are more effective than theoretical and experimental methods; this book aims at the development of related numerical methods, models, and cases for conducting investigations on some scientific and engineering issues. The following will focus on the related research background and current situation of deflection of fracturing fracture network disturbed by discontinuity and multiple fractures in rock, as well as the increasing number of relevant studies in recent years. This chapter will not introduce too much reference information, but will summarise the relevant research progress in the following chapters of each subtopic.

Y. Wang, *Deflection of Fracturing Fracture Network Disturbed by Discontinuity and Multiple Fractures in Rock*, https://doi.org/10.1007/978-981-95-1569-1_1

1.2 Deflection of Fracturing Fracture Network Disturbed by Discontinuity in Rock

(1) Penetration and intersection propagation of fracturing fractures meeting beddings in rock

The penetration and intersection propagation of fracturing fractures meeting beddings with different bedding orientations in rock are shown in Fig. 1.1 (Zhang et al., 2022). Some experimental results indicate that bedding generally affects the propagation mechanism of fractures, and transforms the fracture mode of isotropic specimens from pure open mode to mixed mode at different levels and angles (Khalili et al., 2023); the deviation of fractures caused by sliding and stacking of bedding interfaces leads to the generation of fracture interference, which in turn affects the propagation of multiple fractures (Wu et al., 2022a). The slip properties of the bedding interface determine whether hydraulic fractures can cross the bedding interface, and the slip type and permeability of the bedding interface can cause changes in the width of hydraulic fractures when they pass through the bedding interface, which leads to a discontinuous distribution of local stress and fluid pressure near the bedding interface (Wu et al., 2022b); when the critical traction ratio between bedding planes and rocks drops below 0.7, the fracture propagation mode will change from crossing to deflection (Zeng et al., 2023); during the interaction process, the larger the approach angle of hydraulic fractures relative to the bedding plane, the stronger the inhibitory effect of bedding planes on fracture propagation (Zhang et al., 2022). Figure 1.2 shows the academic publications of penetration and intersection propagation of fracturing fractures meeting beddings in rock in database of Web of Science. It can be seen that since 2010, the number of related papers has increased sharply. In recent years, nearly 30 papers have been published every year; it shows that the research of penetration and intersection propagation of fracturing fractures meeting beddings in rock is concerned and used. It should be noted that the number of published papers in this chapter is in 2023, and some of the papers in that year inevitably have not yet been included in the statistics; therefore, there will be some increase in the actual number of papers in 2023.

(2) Penetration and deflection of fracturing fractures meeting granules in rock

The penetration and deflection of fracturing fractures meeting granules in rock are shown in Fig. 1.3. The hydraulic fractures mainly display four situations when meeting granules: propagation through granules, propagation around granules, stopping within granules, and fracture bifurcation (Yu et al., 2023). The experiments on sand containing granules show that granules with high strength can locally affect the propagation path of hydraulic fractures (Shi et al., 2023); the micro mechanism of the interaction between hydraulic fractures and granules is mainly affected by the uneven distribution of stress field caused by the coordinated deformation of rock mass containing granules (Huang et al., 2023a). The fracture trajectories near material non-uniformity are evaluated to discover fracture arrest and fracture deflection zones near rigid granules (Brescakovic & Kolednik, 2023); experiments have also

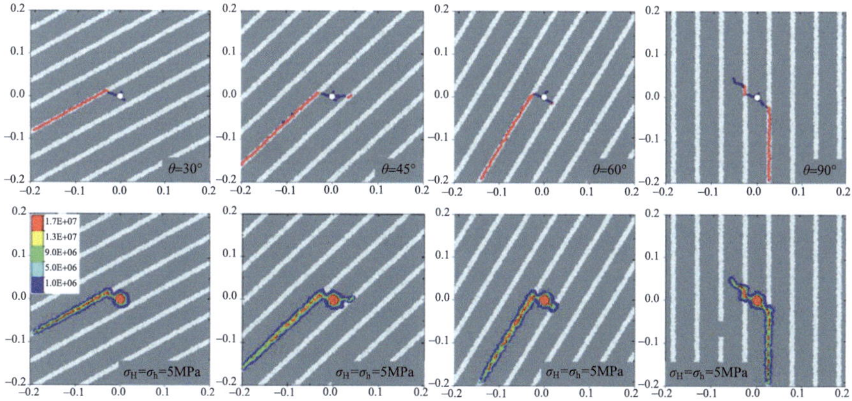

Fig. 1.1 Penetration and intersection propagation of fracturing fractures meeting beddings with different bedding orientations in rock (Zhang et al., 2022)

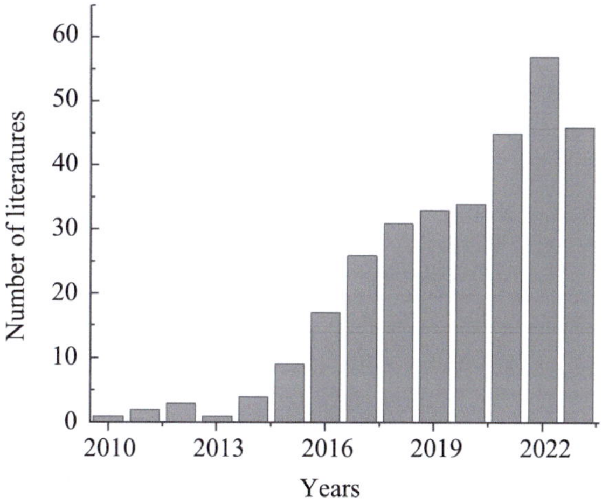

Fig. 1.2 Number of literatures on penetration and intersection propagation of fracturing fractures meeting beddings in rock

shown that the larger the acute angle between the granule direction and the propagation direction of hydraulic fractures, the more conducive it is to the development of penetrating gravel fractures (Li et al., 2022). Figure 1.4 shows the academic publications of penetration and intersection propagation of fracturing fractures meeting granules in rock in database of Web of Science. It can be seen that since 2010, the number of related papers has increased sharply. In recent years, nearly 60 papers have been published every year.

(3) Intersection of fracturing fractures meeting natural fractures in rock

The intersection of fracturing fractures meeting natural fractures in rock is shown in Fig. 1.5, which provides the intersection behavior and fracture width under different

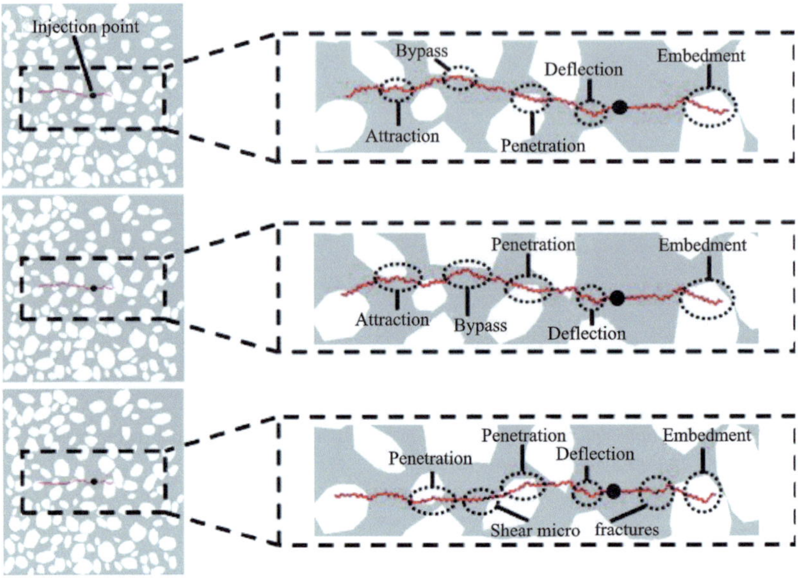

Fig. 1.3 Penetration and deflection of fracturing fractures meeting granules in rock, under different in-situ stress differences: **a** 5 MPa, **b** 10 MPa, and **c** 15 MPa (Huang et al., 2023a)

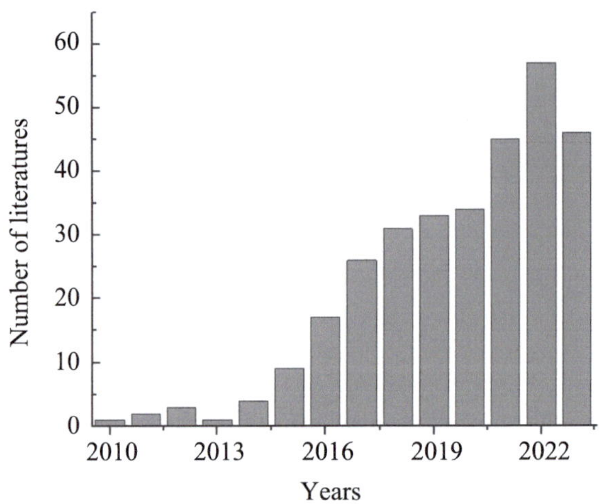

Fig. 1.4 Number of literatures on penetration and intersection propagation of fracturing fractures meeting granules in rock

approaching angles. The experiments found that the larger the stress difference, the less susceptible the propagation path of hydraulic fractures is to the influence of natural fractures, and the more singular the shape of hydraulic fractures, making it less likely to form branches (Xiong & Ma, 2022); compared to hydraulic fracturing parallel or perpendicular to natural fractures, the inclination of natural fractures is conducive to the formation of shear fractures and will deflect the direction of hydraulic fracturing expansion (Qiu et al., 2024). For the propagation of hydraulic fractures under viscosity dominated conditions, hydraulic fractures tend to pass through natural fractures under conditions of short half-length, high injection rate, high fracturing fluid viscosity, high cohesion, high friction coefficient, and high fracture toughness (Liu et al., 2023); the existence of dual parallel natural fractures hinders the propagation of hydraulic fractures along their preferred original path, due to the greater and redistributed leakage through natural fractures, resulting in reduced pressure (Hu et al., 2023). Long natural fractures with appropriate connectivity are more likely to be activated and promote the propagation of hydraulic fractures to far-field areas (Wang et al., 2023). Figure 1.6 shows the academic publications of penetration and intersection propagation of fracturing fractures meeting natural fractures in rock in database of Web of Science. It can be seen that since 2010, the number of related papers has increased sharply. In recent years, nearly 300 papers have been published every year. Compared to the influences of beddings and granules, the disturbance of natural fractures has been given more attention and is considered an important factor affecting the propagation and intersection behaviors of fracturing fractures.

1.3 Deflection of Fracturing Fracture Network Disturbed by Multiple Fractures in Rock

(1) Deflection of fracturing fractures and stress disturbance of multiple perforation clusters in horizontal well

Horizontal wells with multiple fractures in multiple perforation clusters are widely used in unconventional low permeability reservoirs (Roussel & Sharma, 2011), and multiple hydraulic fractures will generate and propagate to form a complex fracture network. However, the interaction between multiple fractures will lead to the deflection of fractures and cause non-planar fractures (Huang et al., 2023b); the reason is that the generation of hydraulic fracture changes the surrounding stress field and causes stress disturbance to the surrounding fractures (Kumar & Ghassemi, 2016; Michael, 2021), which causes the stress shadow effect. The stress shadow effect and the deflection behavior of multiple fractures disturbance have become important factors affecting the fracture network morphology and fracturing efficiency (Sobhaniaragh et al., 2018); the study shows that fracture interaction and stress disturbance between multiple fractures mainly depend on the perforation cluster spacing, fracturing sequence (mainly including sequential fracturing, simultaneous fracturing, and alternate fracturing) (Kumar & Ghassemi, 2016; Liu et al., 2020; Roussel & Sharma, 2011). Fracture deflection under different perforation cluster spacings is

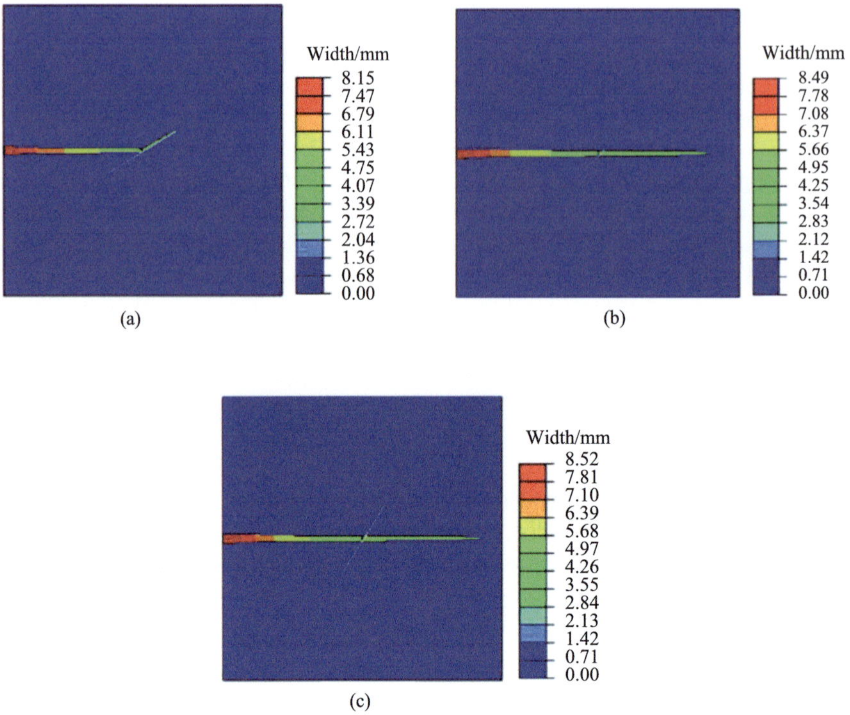

Fig. 1.5 Intersection of fracturing fractures meeting natural fractures in rock, under different approaching angles β: **a** $\beta = 30°$; **b** $\beta = 45°$; **c** $\beta = 60°$ (Sun et al., 2022)

Fig. 1.6 Number of literatures on penetration and intersection propagation of fracturing fractures meeting natural fractures in rock

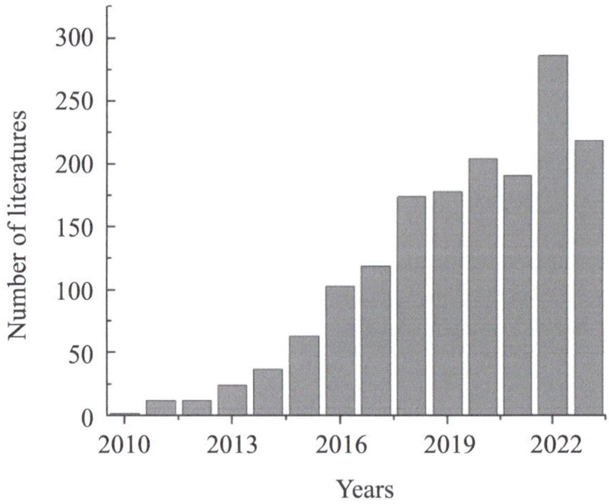

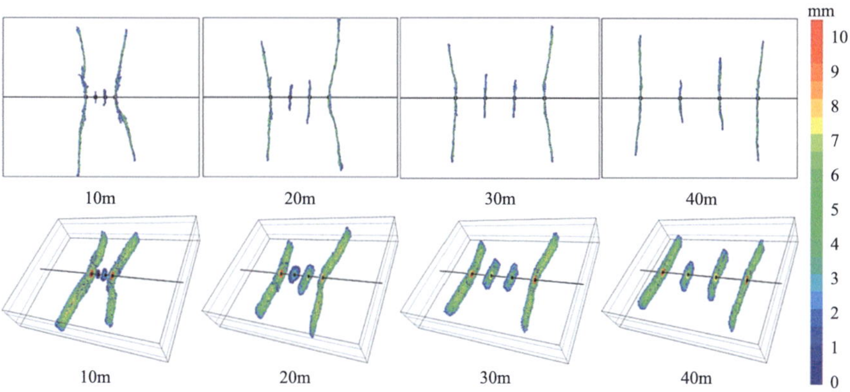

Fig. 1.7 Fracture network morphology under different perforation cluster spacings (the top is the vertical view and the bottom is the three-dimensional view) (Huang et al., 2023b)

shown in Fig. 1.7, with the increase of perforation cluster spacing, the fracture deflection angle decreases, the compressive stress concentration area gradually separates and the stress shadow decreases (Liu et al., 2020); Reasonable perforation cluster spacing can make the fracture propagate fully (He et al., 2020). Fracture network morphology and stress distribution under different fracturing sequences are shown in Fig. 1.8; the stress disturbance and fracture deflection of the subsequent fractures of sequential fracturing accumulates in turn, and the stress disturbance on the intermediate fracture of simultaneous fracturing is greater. In addition to perforation cluster spacing and fracturing sequence, injection rate and bedding dip angle also affect fracture network morphology (Saber et al., 2023; Yang et al., 2018). Figure 1.9 shows the academic publications of deflection of fracturing fractures and stress disturbance of multiple perforation clusters in horizontal well in database of Web of Science. It can be seen that since 2010, the number of related papers has increased sharply. In recent years, nearly 40 papers have been published every year.

(2) Frac-hits and stress disturbance of multi-well fracturing fracture network

Multi-well hydrofracturing will form a more complex fracture network than single-well hydrofracturing, and significantly improve fracturing efficiency and reduce costs (Chen et al., 2018); in addition to the interaction between multiple fractures in the same well, stress disturbance will also occur between multiple wells, resulting in the frac-hits and fracture interaction between multiple wells (He et al., 2020; Wang & Liu, 2023). Both perforation cluster spacing, fracturing sequences and well spacing will affect the stress disturbance between multiple wells and affect the fracture network morphology (Liu et al., 2020). Fracture network morphology of different well spacings as shown in Fig. 1.10; the decrease of well spacing will promote the connection of multi-well fractures (Wang & Liu, 2022; Wu et al., 2018), and too small well spacing will aggravate the interaction between multi-well fractures and lead to strong inhibition of internal fractures (Chen et al., 2018; Liu et al., 2020). The increase of well

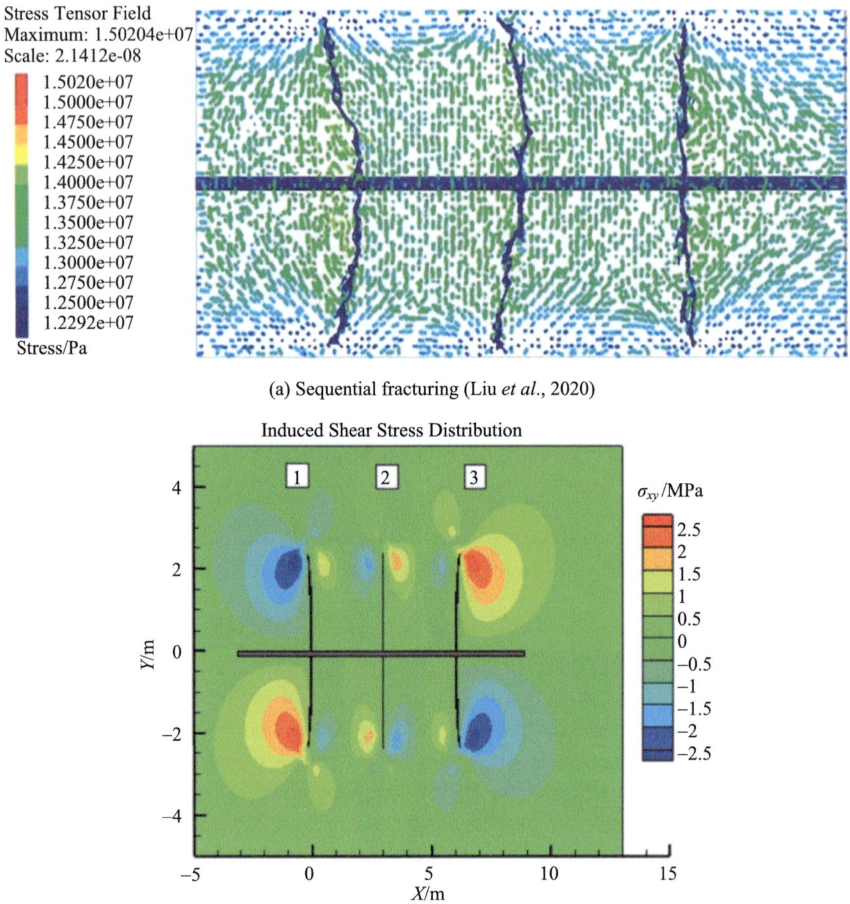

(a) Sequential fracturing (Liu *et al.*, 2020)

(b) Simultaneous fracturing (Kumar and Ghassemi, 2016)

Fig. 1.8 Fracture network morphology and stress distribution under different fracturing sequences

spacing can promote the lateral propagation of interior three-dimensional hydraulic fractures (Chen et al., 2018); therefore, well spacing should be reasonably controlled (Chen et al., 2018). Figure 1.11 shows the fracture network morphology of different fracturing sequences in multiple wells; compared with different fracturing sequences of multiple wells (sequential fracturing, simultaneous fracturing, alternate fracturing and zipper fracturing), the zipper fracturing can avoid frac-hits and fracture connections (Wu et al., 2018), and generate more complex fracture network (Kumar & Ghassemi, 2016). Figure 1.12 shows the academic publications of frac-hits and stress disturbance of multi-well fracturing fracture network in database of Web of Science. It can be seen that since 2010, the number of related papers has increased sharply. In recent years, nearly 35 papers have been published every year.

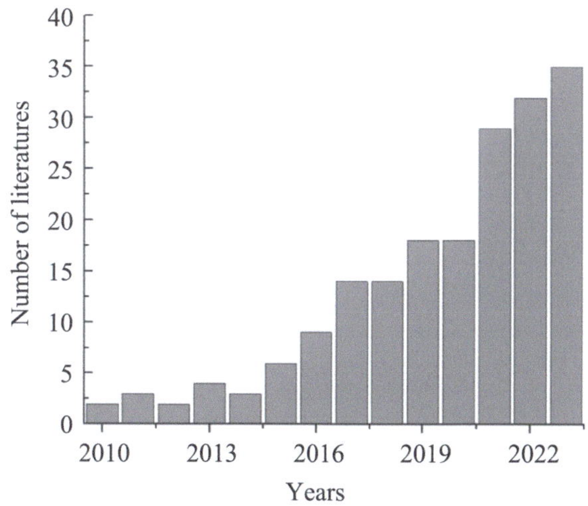

Fig. 1.9 Number of literatures on deflection of fracturing fractures and stress disturbance of multiple perforation clusters in horizontal well

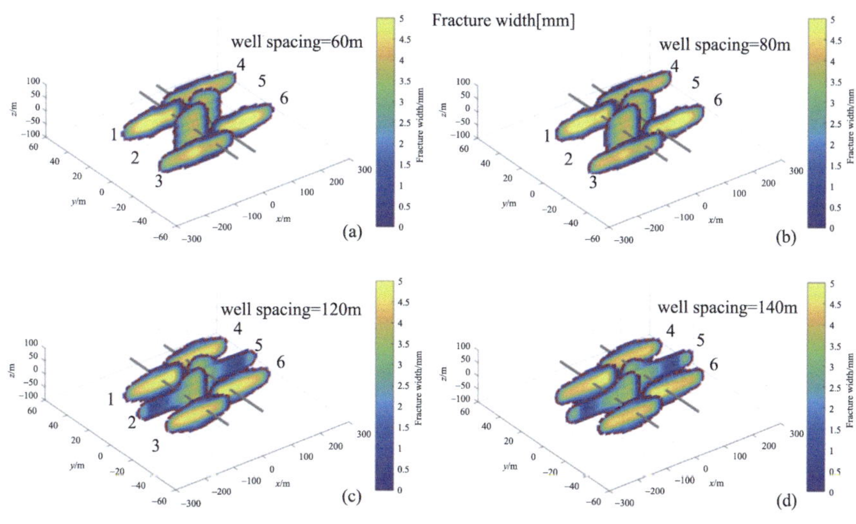

Fig. 1.10 Fracture network morphology under different well spacings (Chen et al., 2018)

1.4 Propagation and Deflection of Fracturing Fracture Network in Supercritical CO_2 Fracturing

Figure 1.13 shows the phase states of CO_2 at different pressures and temperatures, as well as the critical point (304.21 K or 31.06 °C, 7.39 MPa) for distinguishing supercritical, gas, solid, and liquid states. Normally, CO_2 is always in a gaseous state

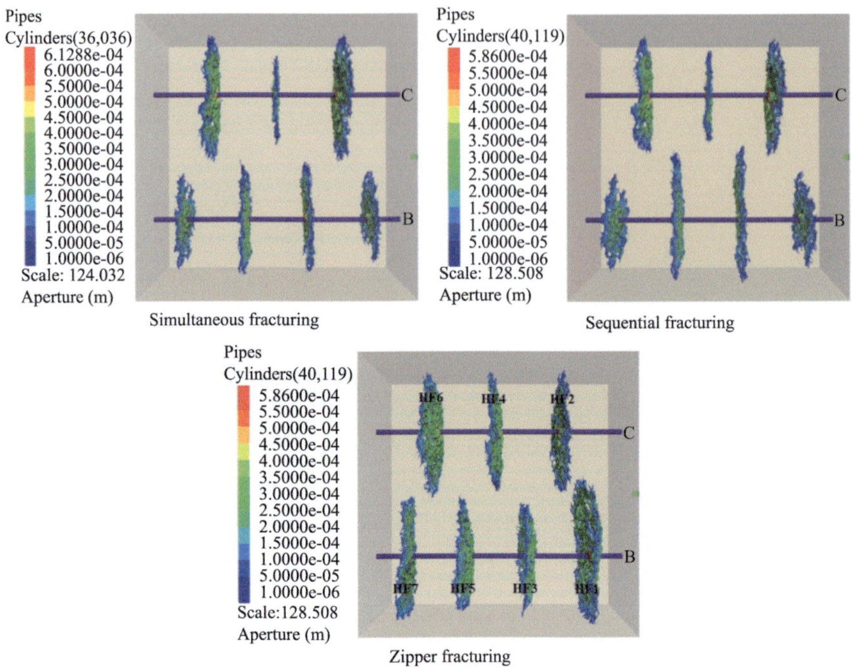

Fig. 1.11 Fracture network morphology under different fracturing sequences in multiple wells (Liu et al., 2020)

Fig. 1.12 Number of literatures on frac-hits and stress disturbance of multi-well fracturing fracture network

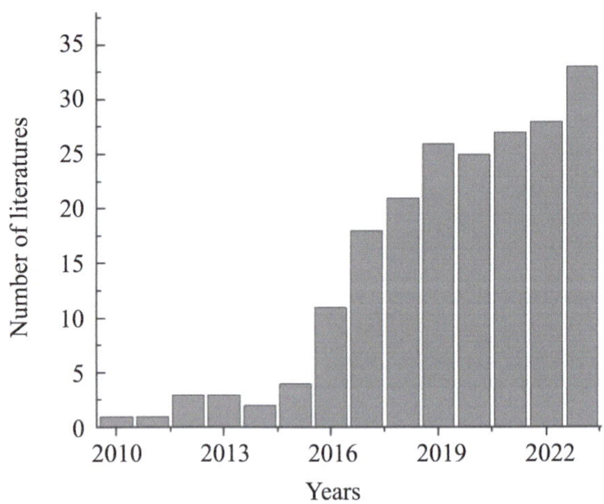

at room temperature and atmospheric pressure; when the temperature and pressure are above the critical point, the phase state of CO_2 gradually transforms into a supercritical state for becoming the SC-CO_2. Compared to water-base fracturing fluids, SC-CO_2 may reduce water cost and, environmental pollution, and has the potential

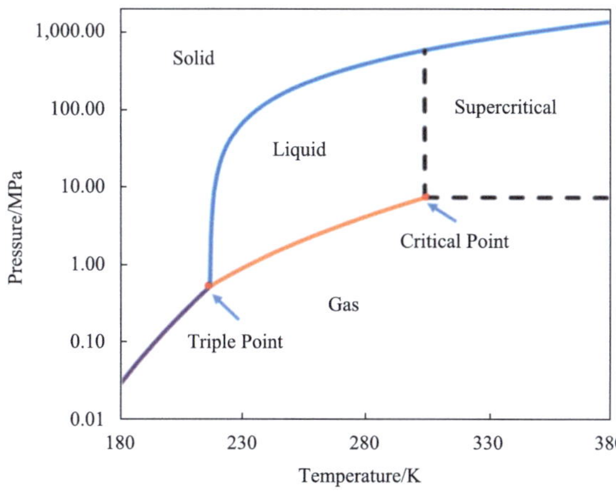

Fig. 1.13 The phase state of CO_2 (Zhao et al., 2021)

for geological storage of CO_2. The performance of SC-CO_2 is gas, but it has a liquid density, and its good properties make it form longer fracturing fracture compared to slick water-driven fracturing fracture as shown in Fig. 1.14. SC-CO_2 has the greater ability to enter micropores and microfractures for improving the permeability of shale matrix (He et al., 2019; Li et al., 2019).

Figure 1.15 shows the experimental and numerical results of water-based fracturing (WBF) and SC-CO_2 fracturing, in which the high dynamic viscosity and low permeability of water-based fracturing fluid make it mainly concentrate in a small area near the fracture, forming a local high-pressure area; however, the dynamic viscosity of SC-CO_2 is relatively low, and the fluid rapidly penetrates into the matrix, forming a wide range of fracturing fluid infiltration areas; many complex branching fractures appear in SC-CO_2 fracturing (Xu et al., 2022). The probability of SC-CO_2 fracturing producing more fractures than that of water-based fracturing to produce well-developed fracture networks (Zhang et al., 2021). The deflection of SC-CO_2 fracturing fracture network is also disturbed by discontinuity in rock. Considering the beddings in rock, the experiments are implemented, and the SC-CO_2 fracturing

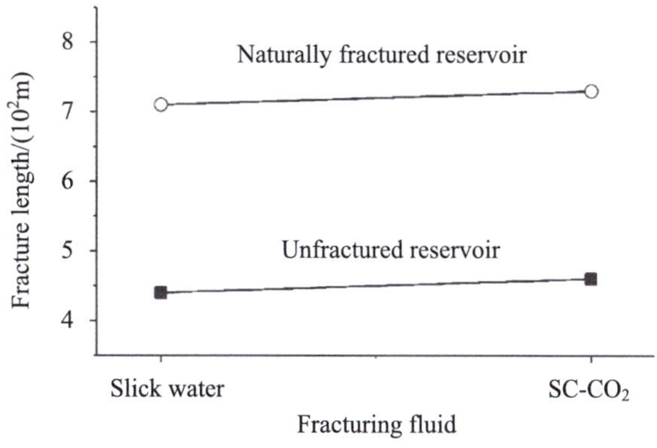

Fig. 1.14 Comparisons of fracture lengths driven by slick water and SC-CO_2 as fracturing fluids (Wang et al., 2019)

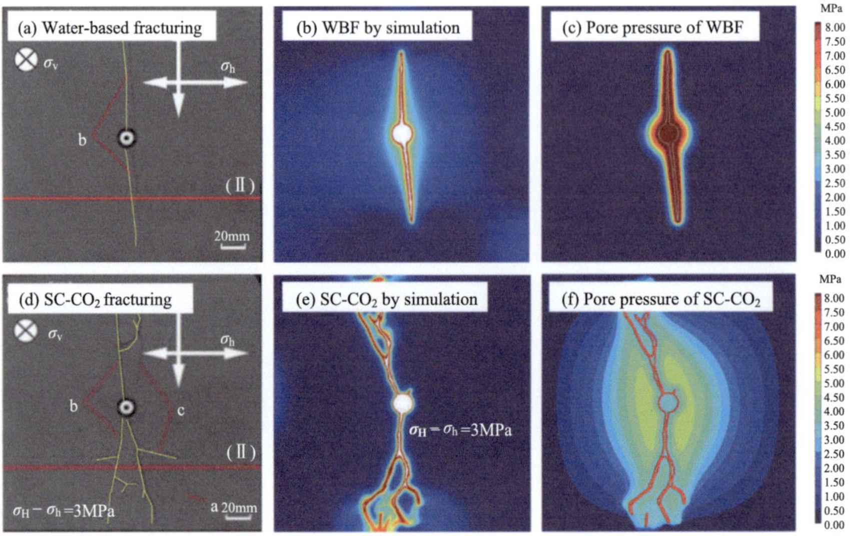

Fig. 1.15 Experimental results of **a** water-based fracturing (WBF) and **d** SC-CO_2 fracturing, numerical results of **b** WBF and **e** SC-CO_2 fracturing, pore pressure distribution of **c** WBF and **f** SC-CO_2 fracturing (Xu et al., 2022)

fracture network is shown in Fig. 1.16; it can be seen that SC-CO_2 fracturing forms a complex branching fracture network, and the fracturing fractures are oriented in different directions intersecting parallel bedding planes. This indicates that the complex SC-CO_2 fracturing fractures are also more likely to contact and connect discontinuous planes (such as the bedding planes, granules, and natural fractures) in reservoir rock for inducing the unusual propagation and deflection of fractures, which makes it difficult to evaluate and optimize the morphology of SC-CO_2 fracturing fracture network (Zhang et al., 2017). Figure 1.17 shows the academic publications on

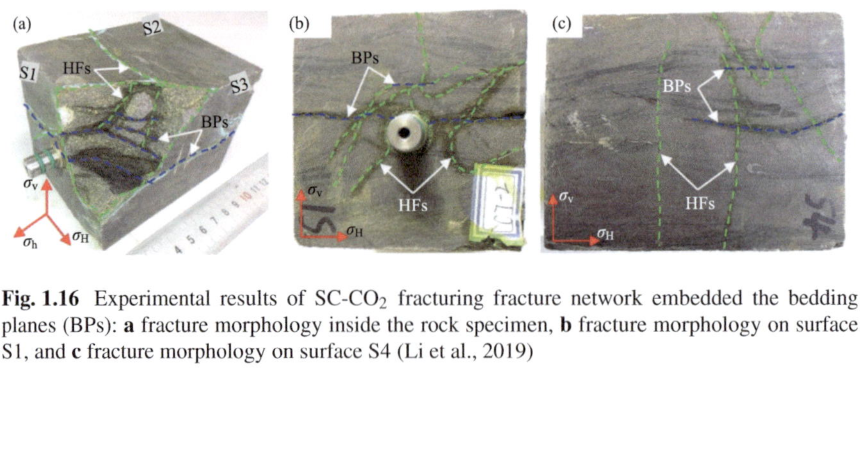

Fig. 1.16 Experimental results of SC-CO_2 fracturing fracture network embedded the bedding planes (BPs): **a** fracture morphology inside the rock specimen, **b** fracture morphology on surface S1, and **c** fracture morphology on surface S4 (Li et al., 2019)

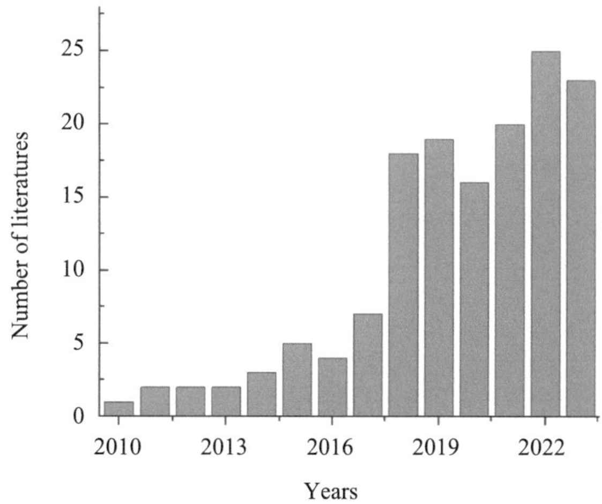

Fig. 1.17 Number of literatures on propagation and deflection of fracturing fracture network in supercritical CO_2 fracturing

propagation and deflection of fracturing fracture network in supercritical CO_2 fracturing in database of Web of Science. It can be seen that since 2010, the number of related papers has increased sharply. In recent years, nearly 25 papers have been published every year.

1.5 Research Contents of the Book

This book implements the numerical investigations on for deflection of fracturing fracture network disturbed by discontinuity and multiple fractures in rock, and some related numerical methods, models and cases are developed. The book covers the following main research contents:

(1) Dynamic intersection and deflection behaviours of hydraulic fractures meeting granules and natural fractures in tight reservoir rock based on statistical modelling and fractal characterization
(2) Deflection behaviours and fractal morphology of hydraulic fractures meeting beddings and granules with variable geometrical configurations and geomechanical properties
(3) Dynamic propagation of tensile and shear fractures induced by impact load in rock based on dual bilinear cohesive zone model
(4) Center- and edge-type intersections of hydraulic fracture network under varying crossed natural fractures and fluid injection rate
(5) Wells connection and long hydraulic fracture induced by multi-well hydrofracturing utilizing crossed perforation clusters
(6) Deflection of fracture networks and gas production in multi-well hydrofracturing utilizing parallel and crossed perforation clusters
(7) Supercritical CO_2-driven intersections of multi-well fracturing fracture network and induced microseismic events in naturally fractured reservoir.

References

Brescakovic, D., & Kolednik, O. (2023). Fracture toughness improvement due to crack deflection and crack trapping by elliptical voids or particles. *International Journal of Solids and Structures, 285*, Article 112551. https://doi.org/10.1016/j.ijsolstr.2023.112551

Chen, X., Li, Y., Zhao, J., Xu, W., & Fu, D. (2018). Numerical investigation for simultaneous growth of hydraulic fractures in multiple horizontal wells. *Journal of Natural Gas Science and Engineering, 51*, 44–52. https://doi.org/10.1016/j.jngse.2017.12.014

He, J., Zhang, Y., Li, X., & Wan, X. (2019). Experimental Investigation on the fractures induced by hydraulic fracturing using freshwater and supercritical CO_2 in shale under uniaxial stress. *Rock Mechanics and Rock Engineering, 52*, 3585–3596. https://doi.org/10.1007/s00603-019-01820-w

He, Y., Yang, Z., Li, X., & Song, R. (2020). Numerical simulation study on three-dimensional fracture propagation of synchronous fracturing. *Energy Science and Engineering, 8*(4), 944–958. https://doi.org/10.1002/ese3.573

Hu, Y., Gan, Q., Hurst, A., & Elsworth, D. (2023). Investigation of coupled hydro-mechanical modelling of hydraulic fracture propagation and interaction with natural fractures. *International Journal of Rock Mechanics and Mining Sciences, 169*, Article 105418. https://doi.org/10.1016/j.ijrmms.2023.105418

Huang, L., He, R., Yang, Z., Tan, P., Chen, W., Li, X., & Cao, A. (2023a). Exploring hydraulic fracture behavior in glutenite formation with strong heterogeneity and variable lithology based on DEM simulation. *Engineering Fracture Mechanics, 278*, Article 109020. https://doi.org/10.1016/j.engfracmech.2022.109020

Huang, L., Tan, J., Fu, H., Liu, J., Chen, X., Liao, X., Wang, X., & Wang, C. (2023b). The non-plane initiation and propagation mechanism of multiple hydraulic fractures in tight reservoirs considering stress shadow effects. *Engineering Fracture Mechanics, 292*, 109570. https://doi.org/10.1016/j.engfracmech.2023.109570.

Khalili, M., Fahimifar, A., & Shobeiri, H. (2023). The effect of bedding planes on the bending strength of rock-like material and evaluation of the crack propagation mechanism. *Theoretical and Applied Fracture Mechanics, 127*, Article 104061. https://doi.org/10.1016/j.tafmec.2023.104061

Kumar, D., & Ghassemi, A. (2016). A three-dimensional analysis of simultaneous and sequential fracturing of horizontal wells. *Geoenergy Science and Engineering, 146*, 1006–1025.

Li, S., Zhang, S., Ma, X., Zou, Y., Li, N., Chen, M., Cao, T., & Bo, Z. (2019). Hydraulic fractures induced by water-/carbon dioxide-based fluids in tight sandstones. *Rock Mechanics and Rock Engineering, 52*, 3323–3340. https://doi.org/10.1007/s00603-019-01777-w

Li, X., Ji, H., Chen, L., Li, M., Xu, K., Jiang, X., Zhang, Z. W., Zhang, Z. H., & Guo, X. (2022). Hydraulic fractures evaluation of the glutenite and the effects of gravel heterogeneity based on cores. *International Journal of Rock Mechanics and Mining Sciences, 160*, Article 105264. https://doi.org/10.1016/j.ijrmms.2022.105264

Liu, X., Rasouli, V., Guo, T., Qu, Z., Sun, Y., & Damjanac, B. (2020). Numerical simulation of stress shadow in multiple cluster hydraulic fracturing in horizontal wells based on lattice modelling. *Engineering Fracture Mechanics, 238*, 107278. https://doi.org/10.1016/j.engfracmech.2020.107278

Liu, T., Wei, X., Liu, X., Liang, L., Wang, X., Chen, J., & Lei, H. (2023). A criterion for a hydraulic fracture crossing a natural fracture in toughness dominant regime and viscosity dominant regime. *Engineering Fracture Mechanics, 289*, Article 109421. https://doi.org/10.1016/j.engfracmech.2023.109421

Michael, A. (2021). Hydraulic fractures from non-uniform perforation cluster spacing in horizontal wells: laboratory testing on transparent gelatin. *Journal of Natural Gas Science and Engineering, 95*, 104158. https://doi.org/10.1016/j.jngse.2021.104158

Qiu, G., Chang, X., Li, J., Guo, Y., Zhou, Z., Wang, L., Wan, Y., & Wang, X. (2024). Study on the interaction between hydraulic fracture and natural fracture under high stress. *Theoretical and Applied Fracture Mechanics, 130*, Article 104259. https://doi.org/10.1016/j.tafmec.2024.104259

Roussel, N. P., & Sharma, M. M. (2011). Optimizing fracture spacing and sequencing in horizontal-well fracturing. *SPE Production & Operations, 26*(2), 173–184. https://doi.org/10.2118/127986-PA

Saber, E., Qu, Q., Sarmadivaleh, M., Aminossadati, S. M., & Chen, Z. (2023). Propagation of multiple hydraulic fractures in a transversely isotropic shale formation. *International Journal of Rock Mechanics and Mining Sciences, 170*, 105510. https://doi.org/10.1016/j.ijrmms.2023.105510

Shi, X., Qin, Y., Gao, Q., Liu, S., Xu, H., & Yu, T. (2023). Experimental study on hydraulic fracture propagation in heterogeneous glutenite rock. *Geoenergy Science and Engineering, 225*, Article 211673. https://doi.org/10.1016/j.geoen.2023.211673

Sobhaniaragh, B., Mansur, W. J., & Peters, F. C. (2018). The role of stress interference in hydraulic fracturing of horizontal wells. *International Journal of Rock Mechanics and Mining Sciences, 106*, 153–164. https://doi.org/10.1016/j.ijrmms.2018.04.024

Sun, T., Zeng, Q., & Xing, H. (2022). A quantitative model to predict hydraulic fracture propagating across cemented natural fracture. *Geoenergy Science and Engineering, 208*(Part C), 109595. https://doi.org/10.1016/j.petrol.2021.109595.

Wang, Y., & Liu, N. (2022). Dynamic propagation and shear stress disturbance of multiple hydraulic fractures: Numerical cases study via multi-well hydrofracturing model with varying adjacent spacings. *Energies, 15*(13), 4621. https://doi.org/10.3390/en15134621

Wang, Y., & Liu, N. (2023). Unstable propagation of hydraulic fractures under varying well spacing and initiation sequence of multiple horizontal wells considering shear stress shadows. *Engineering Computations, 40*(9/10), 2483–2509. https://doi.org/10.1108/EC-03-2023-0096

Wang, Y., Ju, Y., Chen, J., & Song, J. (2019). Adaptive finite element-discrete element analysis for the multistage supercritical CO_2 fracturing and microseismic modelling of horizontal wells in tight reservoirs considering pre-existing fractures and thermal-hydromechanical coupling. *Journal of Natural Gas Science and Engineering, 61*, 251–269. https://doi.org/10.1016/j.jngse.2018.11.022

Wang, J., Xie, H., Matthai, S. K., Hu, J., & Li, C. (2023). The role of natural fracture activation in hydraulic fracturing for deep unconventional geo-energy reservoir stimulation. *Petroleum Science, 20*(4), 2141–2164. https://doi.org/10.1016/j.petsci.2023.01.007

Wu, K., Wu, B., & Yu, W. (2018). Mechanism analysis of well interference in unconventional reservoirs: Insights from fracture-geometry simulation between two horizontal wells. *SPE Production and Operations, 33*(1), 12–20. https://doi.org/10.2118/186091-PA

Wu, S., Gao, K., Feng, Y., & Huang, X. (2022a). Influence of slip and permeability of bedding interface on hydraulic fracturing: A numerical study using combined finite-discrete element method. *Computers and Geotechnics, 148*, Article 104801. https://doi.org/10.1016/j.compgeo.2022.104801

Wu, S., Gao, K., Wang, X., Ge, H., Zou, Y., & Zhang, X. (2022b). Investigating the propagation of multiple hydraulic fractures in shale oil rocks using acoustic emission. *Rock Mechanics and Rock Engineering, 55*(10), 6015–6032. https://doi.org/10.1007/S00603-022-02960-2

Xiong, D., & Ma, X. (2022). Influence of natural fractures on hydraulic fracture propagation behavior. *Engineering Fracture Mechanics, 276*(Part A), 108932. https://doi.org/10.1016/j.engfracmech.2022.108932.

Xu, W., Yu, H., Zhang, J., Lyu, C., Wang, Q., Micheal, M., & Wu, H. (2022). Phase-field method of crack branching during SC-CO_2 fracturing: a new energy release rate criterion coupling pore pressure gradient. *Computer Methods in Applied Mechanics and Engineering, 399*, 115366. https://doi.org/10.1016/j.cma.2022.115366.

Yang, Z. Z., Yi, L. P., Li, X. G., & He, W. (2018). Pseudo-three-dimensional numerical model and investigation of multi-cluster fracturing within a stage in a horizontal well. *Geoenergy Science and Engineering, 162*, 190–213. https://doi.org/10.1016/j.petrol.2017.12.034

Yu, S., Zhou, Y., Yang, J., & Chen, W. (2023). Hydraulic fracturing modelling of glutenite formations using an improved form of SPH method. *Geoenergy Science and Engineering, 227*, Article 211842. https://doi.org/10.1016/j.geoen.2023.211842

Zeng, Q., Bo, L., Li, Q., Sun, J., & Yao, J. (2023). Numerical investigation of hydraulic fracture propagation interacting with bedding planes. *Engineering Fracture Mechanics, 291*, Article 109575. https://doi.org/10.1016/j.engfracmech.2023.109575

Zhang, X., Lu, Y., Tang, J., Zhou, Z., & Liao, Y. (2017). Experimental study on fracture initiation and propagation in shale using supercritical carbon dioxide fracturing. *Fuel, 190*, 370–378. https://doi.org/10.1016/j.fuel.2016.10.120

Zhang, C., Cheng, P., Ma, Z., Ranjith, P., & Zhou, J. (2021). Comparison of fracturing unconventional gas reservoirs using CO_2 and water: an experimental study. *Journal of Petroleum Science and Engineering, 203*, 108598. https://doi.org/10.1016/j.petrol.2021.108598.

Zhang, Y. L., Liu, Z., Han, B., Zhu, S., & Zhang, X. (2022). Numerical study of hydraulic fracture propagation in inherently laminated rocks accounting for bedding plane properties. *Geoenergy Science and Engineering, 210*, Article 109798. https://doi.org/10.1016/j.petrol.2021.109798

Zhao, H., Wu, K., Huang, Z., Xu, Z., Shi, H., & Wang, H. (2021). Numerical model of CO_2 fracturing in naturally fractured reservoirs. *Engineering Fracture Mechanics, 244*, 107548. https://doi.org/10.1016/j.engfraemech.2021.107548.

Chapter 2
Dynamic Intersection and Deflection Behaviours of Hydraulic Fractures Meeting Granules and Natural Fractures in Tight Reservoir Rock Based on Statistical Modelling and Fractal Characterization

2.1 Introduction

The morphology of hydraulic fractures is mainly affected by deflection behaviours of fractures meeting embedded heterogeneous and discontinuous geological structures, such as the granules and natural fractures. The intersection and deflection behaviours (penetration, diversion, and arrest) of hydraulic fractures arise in heterogeneous reservoir embedded granules and natural fractures, as shown in Fig. 2.1. Evaluating these behaviours and morphology of hydraulic fractures is a key scientific issue to control and optimize the fracturing effects. It is urgent to quantitatively investigate the intersection and deflection behaviours and quantitative morphology of hydraulic fractures in heterogeneous rock masses and analyze the influences of heterogeneity on fracturing effects.

For gravels in rock, their percentage content, geometrical size, and geomechanical properties may induce the termination, deflection, branching, and penetration behaviours of fractures (Hou et al., 2017; Li et al., 2013; Rui et al., 2018), and the fracture morphology is affected by the heterogeneity of gravels (Li et al., 2018; Sharafisafa et al., 2020). The large-scale triaxial hydrofracturing system is used to quantify the influence of gravel on fracture propagation considering different in-situ stress differences and gravel sizes (Ma et al., 2017); the stronger the non-uniformity of granule size, the smaller the uniaxial compressive strength of rock mass (Liu et al., 2018). The micro-heterogeneity of granule size may control the macroscopic response and mechanical microscopic behaviour of rocks (Lan et al., 2010; Mngadi et al., 2019; Wang & Liu, 2021); the macroscopic fractures may penetrate the gravels once they meet fractures (Xv et al., 2019; Yan et al., 2019). For natural fractures in rock, when the hydraulic fractures mee natural fractures, the deflection and penetration behaviours of hydraulic fractures may arise (Warpinski, 1991; Zhang et al., 2018a, b). The opening or closing of natural fractures may affect fracture penetration

Y. Wang, *Deflection of Fracturing Fracture Network Disturbed by Discontinuity and Multiple Fractures in Rock*, https://doi.org/10.1007/978-981-95-1569-1_2

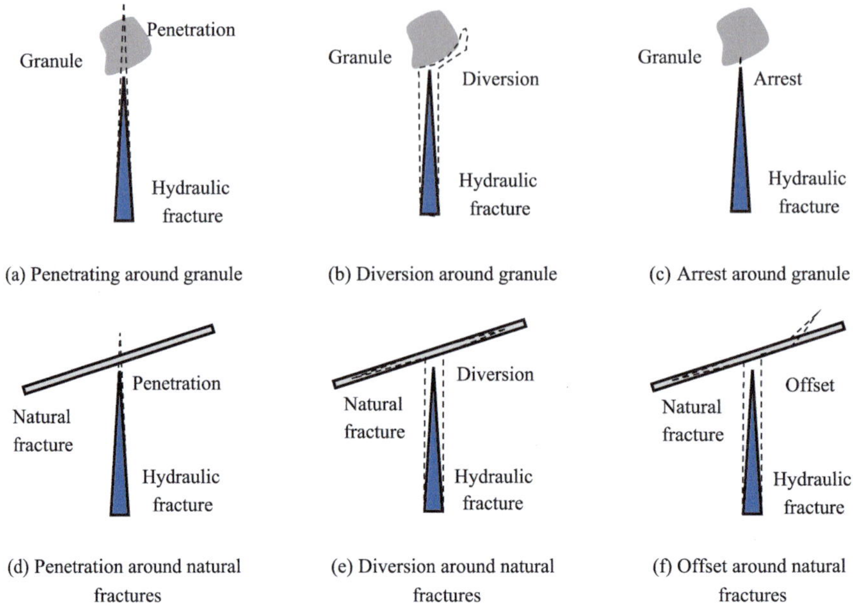

Fig. 2.1 Schematic diagram of hydraulic fracture propagation in a heterogeneous reservoir embedded granules and natural fractures

(Daneshy, 1974), and the surface friction and normal stress acting on natural fractures also affect hydraulic fracture propagation (Anderson, 1981). Some significant factors in the propagation process of hydraulic fractures, such as the high stress difference, approach angle between fractures (Blanton, 1982), stress distribution (Blanton, 1986; Olson et al., 2012), will affect the trajectory of fracture networks. The geometrical shapes and structural characteristics of natural fractures will lead to changes of deflection and penetration behaviours of hydraulic fractures (Wang et al., 2014, 2017, 2021). Therefore, the morphology of hydraulic fracture networks is influenced by the heterogeneous geometrical characteristics of reservoirs, and targeted research on fracture propagation laws and mechanisms under the heterogeneous geometrical characteristics of reservoir rock masses is urgent.

To investigate the evolution process of hydrofracturing fractures, the field detection technologies (Bunger et al., 2015; Warpinski et al., 2009) and theoretical models are developed (Geertsma & De Klerk, 1969; Khristianovich and Zheltov, 1955). However, it is challenging for field and theoretical models to capture the dynamic propagation process of hydrofracturing fracture and the complex evolution process of involved physical fields. Therefore, in order to simulate the complex propagation behaviour of fractures in fluid-driven fracture propagation, many numerical simulation methods are proposed. The finite element method is a commonly used numerical method for conducting response analysis of rock mechanics (Biot, 1955), but due to its difficulty in achieving fracture propagation, it has exposed problems in hydraulic fracture research, requiring the development of some novel numerical methods and models (Gomaa et al., 2014). Furthermore, some numerical methods are proposed, such as the extended finite element method, discrete element method, discrete fracture network method, boundary element method, combined finite element-discrete element method, and finite element-discrete element-finite volume method, are proposed (Wang & Zhang, 2022), and some the parallel computation techniques are used to analyse the hydrofracturing behaviours of reservoirs (Wang et al., 2022). The combined finite element and discrete element methods can be used for continuum fields evolution and discontinuum fracture field analysis in hydrofracturing processes; based on this method, this study will also establish models and conduct continuum-discontinuous analysis.

To derive the influences of reservoir heterogeneity on hydraulic fracture propagation by numerical methods, heterogeneous factors such as granules and natural fractures need to be considered when constructing numerical models. As digital image processing (DIP) technology is used as a powerful tool to explore the microstructure of materials and the spatial distribution of minerals (Ali et al., 2024), combining DIP images into mechanical modelling to generate numerical samples with microstructure extracted from rock slice images has become an important method to establish models (Yue et al., 2003). The high-resolution microcomputer tomography technology is applied to the true triaxial physics experiment of hydrofracturing in heterogeneous sandstone, and the visual description of fracture propagation is obtained; DIP technology can simulate the actual micro geometrical heterogeneity and microstructure of rocks in numerical simulation, but it requires a large number of representative rock slice images to ensure accuracy. Another strategy considering the influence of micro heterogeneity in the numerical model is to use statistical parameters to include heterogeneity in the numerical model (Liu et al., 2006; Manouchehrian & Cai, 2016; Mousavi et al., 2018), which methods discuss the influences of heterogeneity through the statistical distribution of material properties or heterogeneity shapes to investigate the influences of material heterogeneity (Li et al., 2020, 2004; Potyondy & Cundall, 2004; Wong et al., 2006). However, the relationships and impact mechanisms between the statistical heterogeneity of reservoir and hydraulic fracturing networks are not yet clear and need to be clarified through research.

Besides, accurate quantitative evaluation of the formation of hydraulic fracture networks is the precondition and basics for control and optimization. Fractal geometry can provide a potential way to quantify disordered and irregular objects, and fracture networks can be quantitative characterized using fractal approaches (Boguna et al., 2021; Ghanbarian & Hunt, 2017). Some researchers found that natural and synthetic fracture networks show fractal characteristics (Berkowitz & Hadad, 1997); by developing the box-counting method, the dynamic fracture network propagation of porous media may be characterized (Cai et al., 2017); furthermore, the multifractal methods and scanning electron microscope images are used to analyze the microscopic pore structure of shale rock (Liu & Ostadhassan, 2017). Based on these methods, the micro-anisotropy and pore structure of coal show typical fractal characteristics (Frosch et al., 2000; Mondal et al., 2017), and the fractal characteristics of adsorption pores are detected (Zhao et al., 2022). The complexity of hydraulic fractures can also be described by fractal dimension, which can be used to evaluate fracture network morphology and fracturing effect under different in-situ and heterogeneous properties (Jiang et al., 2020; Movassagh et al., 2021). Fractal dimension analysis has become a potential method for characterizing the complex morphology of hydraulic fractures.

This chapter is organized as follows: Sect. 2.2 introduces the statistical modelling for tight heterogeneous reservoir rock, including statistical uniform and Weibull distribution of heterogeneous reservoir rock and establishment process of statistical models with granules and natural fractures. Section 2.3 introduces the governing partial differential equations and numerical discretization of hydrofracturing in fractured porous media. Section 2.4 introduces the fractal characterization method for fracture network morphology. Section 2.5 introduces the global procedure for statistical modelling, fracture propagation, and fractal characterization. Section 2.6 introduces the results and discussions. Section 2.7 summarizes the conclusions.

2.2 Statistical Modelling for Tight Heterogeneous Reservoir Rock

2.2.1 Statistical Uniform and Weibull Distribution of Heterogeneous Reservoir Rock

This study discusses the influences of heterogeneity of reservoir rock mass on dynamic intersection and deflection behaviours of hydraulic fractures, especially the influences of the geometrical distribution of heterogeneous granules and natural fractures. In this study, the geometrical location information of heterogeneous factors is in the form of uniform distribution, which is used to simulate the statistical uniform distribution of heterogeneous granules and natural fractures in rock mass; the geometrical size information of heterogeneous factors is in the form of Weibull distribution, which is used to simulate the statistical Weibull distribution of heterogeneous granule size and natural fracture length in rock mass.

The uniform distribution is used to describe the distribution of location parameters of granules and natural fractures, and the probability density function $f(x)$ and distribution function $F(x)$ are as follows:

$$f(x) = \begin{cases} 0 & \text{other} \\ \frac{1}{n-m} & m < x < n \end{cases} \tag{2.1}$$

$$F(x) = \begin{cases} 0 & x < m \\ \frac{x-m}{n-m} & m \le x \le n \\ 1 & x > n \end{cases} \tag{2.2}$$

where x is the distribution parameter of location variable, which has the range of $m \le x \le n$. The functions $f(x)$ and $F(x)$ are shown in Fig. 2.2.

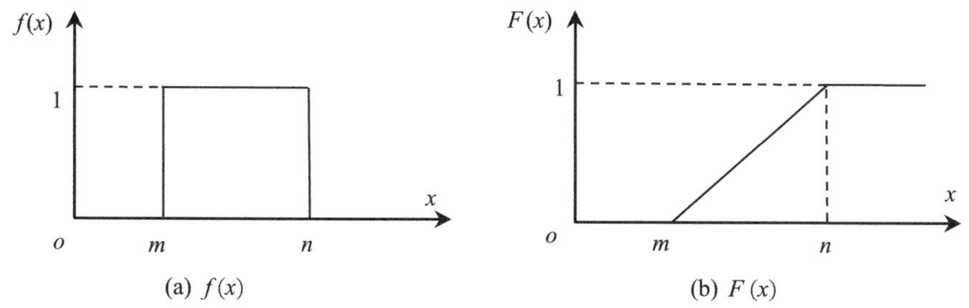

Fig. 2.2 Probability density function $f(x)$ and distribution function $F(x)$ of statistical uniform distribution

The Weibull distribution is used to describe the random distribution of geometrical parameters of granules and natural fractures, and the probability density function $f(\lambda; \lambda_0; k)$ and distribution function $F(\lambda; \lambda_0; k)$ are as follows:

$$f(\lambda; \lambda_0; k) = \begin{cases} 0 & \lambda < 0 \\ \frac{k}{\lambda_0}\left(\frac{\lambda}{\lambda_0}\right)^{k-1} \mathrm{e}^{-(\lambda/\lambda_0)^k} & \lambda \geq 0 \end{cases} \tag{2.3}$$

$$F(\lambda; \lambda_0; k) = \begin{cases} 0 & \lambda < 0 \\ 1 - \mathrm{e}^{-(\lambda/\lambda_0)^k} & \lambda \geq 0 \end{cases} \tag{2.4}$$

where λ is the distribution parameter of heterogeneous components; λ_0 is the average value of distribution parameter; k is the concentration degree of distribution. The functions $f(\lambda; \lambda_0; k)$ and $F(\lambda; \lambda_0; k)$ are shown in Fig. 2.3, in which the influences of variable k and variable λ_0 on the functions are detected: The smaller the concentration degree k is, the flatter the function $f(\lambda; \lambda_0; k)$ is, and the wider the distribution range of heterogeneous parameter (granule diameter or natural fracture length) is; the larger the concentration degree k is, the more compact the function $f(\lambda; \lambda_0; k)$ is, and the narrower the distribution range of heterogeneous parameter (granule diameter or natural fracture length) is. In this study, the Weibull distribution parameters for numerical cases of reservoirs embedded granules and natural fractures are shown in Tables 2.1 and 2.2.

2.2.2 *Establishment Process of Statistical Models with Granules and Natural Fractures*

The geometrical models of heterogeneous tight reservoirs embedded granules and natural fractures are shown in Fig. 2.4: The model contains heterogeneous granules and natural fractures; in-situ stresses are applied at the boundaries of the model; fracturing fluid is injected in the middle. Based on the statistical uniform and Weibull distribution of heterogeneous reservoir rock, the establishment process of statistical models with granules and natural fractures is implemented:

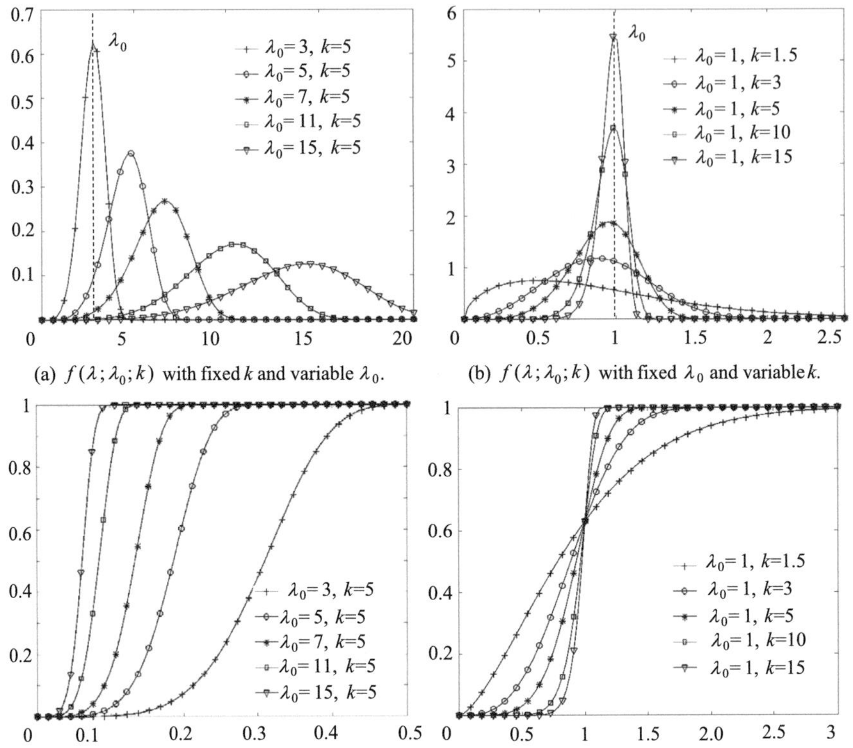

(a) $f(\lambda;\lambda_0;k)$ with fixed k and variable λ_0.

(b) $f(\lambda;\lambda_0;k)$ with fixed λ_0 and variable k.

(c) $F(\lambda;\lambda_0;k)$ with fixed k and variable λ_0.

(d) $F(\lambda;\lambda_0;k)$ with fixed λ_0 and variable k.

Fig. 2.3 Probability density function $f(\lambda;\ \lambda_0;\ k)$ and distribution function $F(\lambda;\ \lambda_0;\ k)$ of statistical Weibull distribution

Table 2.1 Weibull distribution parameters for numerical cases of reservoirs embedded granules

Cases	λ_0/cm	k
Granule-I	3	15
Granule-II	5	15
Granule-III	7	15

Table 2.2 Weibull distribution parameters for numerical cases of reservoirs embedded natural fractures

Cases	λ_0/cm	k
Fracture-I	4	15
Fracture-II	8	15
Fracture-III	12	15

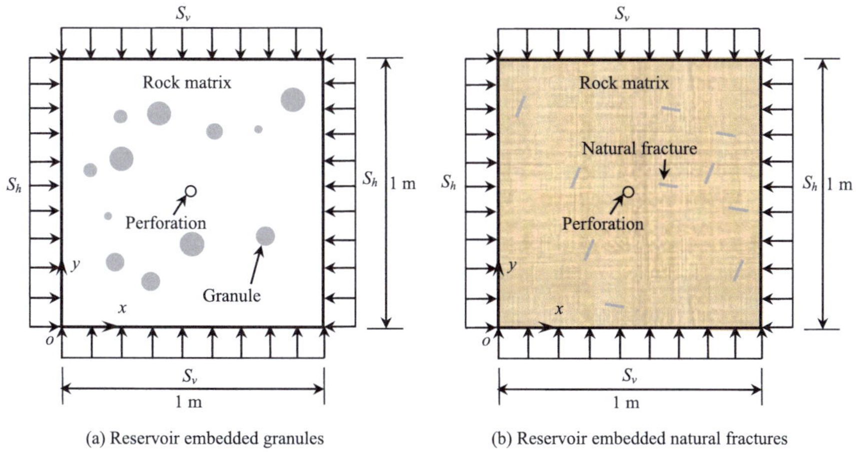

Fig. 2.4 Geometrical models of heterogeneous tight reservoirs embedded granules and natural fractures

(a) To discuss the influences of geometrical distribution of heterogeneous granules, the granule shape is assumed to be circular, and the center of the circular granule is specifically represented by each coordinates following the uniform distribution. The granule diameter follows the Weibull distribution with the average value of distribution parameter λ_0 and the concentration degree k. The Weibull distribution parameters for numerical cases (Granule-I, Granule-II, and Granule-III) of reservoirs embedded granules are shown in Table 2.5, and the corresponding statistical geometrical models of heterogeneous tight reservoirs embedded granules are shown in Fig. 2.5.

(b) The main factors affecting the geometrical distribution law of natural fractures in rock mass are fracture length, dip angle, connectivity, opening, density, etc. Based on the changes of these parameters, the seepage characteristics of rock will be affected. In this study, the geometrical location, dip angle, and length of fractures are modelled and analyzed. The natural fracture shape is assumed to be slender rectangle, and the center of the slender natural fracture is specifically represented by each coordinate following the uniform distribution; the dip angle of natural fracture obeys the uniform distribution of $[0, 2\pi]$. The natural fracture length follows the Weibull distribution with the average value of distribution parameter λ_0 and the concentration degree k. The Weibull distribution parameters for numerical cases (Fracture-I, Fracture-II, and Fracture-III) of reservoirs embedded natural fractures are shown in Table 2.5, and the corresponding statistical geometrical models of heterogeneous tight reservoirs embedded natural fractures are shown in Fig. 2.6.

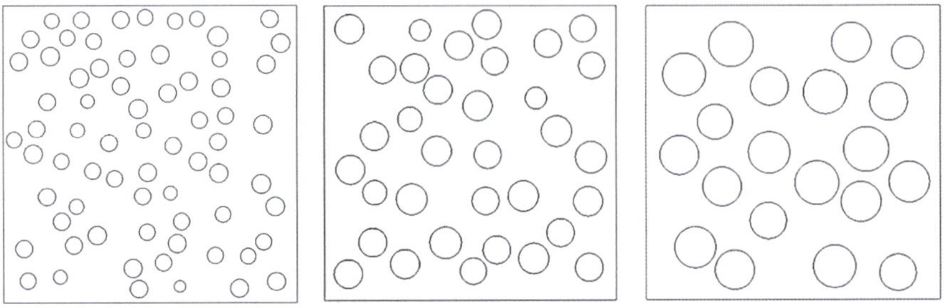

(a) Case Granule-I: $\lambda_0 = 3$, k=15

(b) Case Granule-II: $\lambda_0 = 5$, $k = 15$

(c) Case Granule-III: $\lambda_0 = 7$, k =15

Fig. 2.5 Statistical geometrical models of heterogeneous tight reservoirs embedded granules

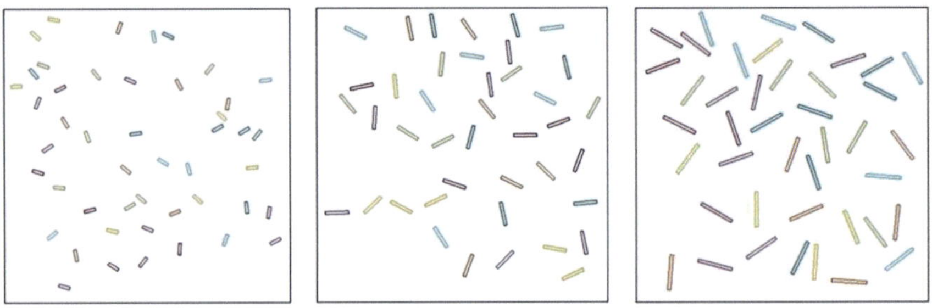

(a) Case Fracture-I: $\lambda_0 = 4$, $k = 15$

(b) Case Fracture-II: $\lambda_0 = 8$, $k = 15$

(c) Case Fracture-III: $\lambda_0 = 12$, $k = 15$

Fig. 2.6 Statistical geometrical models of heterogeneous tight reservoirs embedded natural fractures

2.3 Governing Partial Differential Equations and Numerical Discretization of Hydrofracturing in Fractured Porous Media

2.3.1 Governing Equation of Solid Deformation

Considering the influence of dynamic inertia, the mechanical equilibrium equation of solid deformation on representative elements in Cartesian coordinates is (Wang & Zhang, 2022; Wang et al., 2022):

$$\nabla \cdot \boldsymbol{\sigma} = \rho \ddot{\boldsymbol{u}} + c\dot{\boldsymbol{u}} - \boldsymbol{f}, \quad x, y, z \in \Omega \tag{2.5}$$

where $\boldsymbol{u}(x, y, z) = (u(x, y, z), v(x, y, z), w(x, y, z),)^{\mathrm{T}}$ is the displacement vector; $\boldsymbol{\sigma}$ is the stress tensor; $\boldsymbol{f}$ is the external load vector (the force parameters of the representative element), including the body force, the fluid pressure on the fracture surface, the spring force, and the interface force; ρ is density; c is the damping coefficient; $\dot{\boldsymbol{u}}$ and $\ddot{\boldsymbol{u}}$ are the velocity vector and the acceleration vector, respectively, which are the derivatives of the displacement vector to time t. Ω is the solution domain.

2.3.2 Governing Equations of Fluid Flow in Fractured Porous Media

The fracture network formed by hydrofracturing and original porous media constitutes fractured porous media. In Eqs. (2.2) and (2.3), the variable containing the subscript m represents the variable in the porous elastic rock matrix, and the variable containing the subscript f represents the variable in the hydraulic fracture. Assuming that the fluid in the fractured porous media is a single-phase flow, the simplified Darcy's law of fluid flow in porous media and hydraulic fractures can be used, and the gravity effect can be ignored to obtain the velocity field (Wang et al., 2022):

$$v_m = -\frac{k_m}{\mu}\nabla p_m, \quad x, y, z \in \Omega \tag{2.6}$$

$$v_f = -\frac{k_f}{\mu}\nabla p_f, \quad x, y, z \in \Omega \tag{2.7}$$

where v_m and v_f are the fluid flow velocity fields of porous media and hydraulic fractures, respectively. p_m and p_f are the pressure values of porous media and hydraulic fractures, respectively. k_m and k_f are the permeability of porous media and hydraulic fractures, respectively; μ is the viscosity coefficient. In the absence of gravity and capillary force, the pressure equation of single-phase flow in porous media and hydraulic fractures can be written as follows:

$$S_m\dot{p}_m - \nabla \cdot V_m = -\alpha\frac{\partial \varepsilon_V}{\partial t}, \quad x, y, z \in \Omega \tag{2.8}$$

$$S_f\dot{p}_f - \nabla \cdot V_f = q_f, \quad x, y, z \in \Omega \tag{2.9}$$

where $S_m = n/K_f$ and $S_f = 1/K_f$ are the water storage coefficients of porous media and hydraulic fractures, respectively. n is porosity, where the porosity of the fracture is equal to 1, q_f is the source of an external fluid, and ε_V is the volume strain of the rock matrix.

2.3.3 Fracture Criterion

The failure, sliding and fracture of solids occur at the interface between elements. The maximum tensile stress criterion and Mohr–Coulomb strength criterion are used to judge the tensile and shear failure of the element:

$$\text{Tensile failure criterion:}\quad \sigma_n = \overline{\sigma}_n \tag{2.10}$$

$$\text{Shear failure criterion:}\quad \tau \geq \sigma_n \tan\varphi + C \tag{2.11}$$

where, σ_n and τ are normal stress and tangential stress respectively; $\overline{\sigma}_n$ is tensile strength; C is cohesion; φ is the angle of internal friction.

For the statistical models with granules and natural fractures, the basic physical parameters of the rock matrix are shown in Table 2.3, and the basic physical parameters of granule and natural fracture members are shown in Table 2.4.

Table 2.3 Basic physical parameters of the rock matrix

Parameters	Value
Horizontal in-situ stress in x direction S_h/MPa	60
Vertical in-situ stress in y direction S_v/MPa	60
Fluid injection rate Q/(m^3/s)	0.0001
Poisson's ratio of rock matrix v	0.1
Young's modulus of rock matrix E/GPa	22.38
Tensile strength of rock matrix $\overline{\sigma}_n$/MPa	8.25
Porosity ϕ/%	0.4
Internal friction angle of rock matrix φ/(°)	22.9
Cohesion of rock matrix C/MPa	14.64
Density ρ_b/(kg/m^3)	2.5×10^3
Permeability k/m^2	1×10^{-16}
Gravity g/(m/s^2)	9.81
Damping coefficient c	0.8
Biot's coefficient α	0.7

Table 2.4 Basic physical parameters of granule and natural fracture members

Granule parameters	Value	Natural fracture parameters	Value
Young's modulus E/GPa	43.88	Young's modulus E/GPa	22.38×10^{-8}
Cohesion C/MPa	52.44	Cohesion C/MPa	52.44×10^{-6}
Tensile strength $\overline{\sigma}_n$/MPa	25	Tensile strength $\overline{\sigma}_n$/MPa	8.25×10^{-5}
Internal friction angle φ/(°)	26.13	Internal friction angle φ/(°)	22.9
Poisson's ratio v	0.08	Poisson's ratio v	0.1

2.3.4 Numerical Discretization Based on the Combined Finite Element-Discrete Element-Finite Volume Method

Using the variational formula to solve the solid deformation control equation, Eq. (2.5) can be transformed into the following matrix form on the element e:

$$\boldsymbol{M}^e\ddot{\boldsymbol{D}}(t) + \boldsymbol{C}^e\dot{\boldsymbol{D}}(t) + \boldsymbol{K}^e\boldsymbol{D}(t) = \boldsymbol{F}^e, \quad x, y, z \in \Omega \tag{2.12}$$

where $\boldsymbol{D}(t)$ is the displacement vector composed of the node displacements of element e, $\boldsymbol{M}^e$, $\boldsymbol{C}^e$ and $\boldsymbol{K}^e$ are the mass, damping and stiffness matrices, $\dot{\boldsymbol{D}}(t)$ and $\ddot{\boldsymbol{D}}(t)$ are the vectors of node velocity and acceleration at time t, and $\boldsymbol{F}^e$ is the external load vector. $\boldsymbol{F}^e$ can be expressed as:

$$\boldsymbol{F}^e = \boldsymbol{F}^e_{\mathrm{b}} + \boldsymbol{F}^e_{\mathrm{p}} + \boldsymbol{F}^e_{\mathrm{s}} + \boldsymbol{F}^e_{\mathrm{t}} \tag{2.13}$$

where $\boldsymbol{F}^e_{\mathrm{b}}$ is the body force, $\boldsymbol{F}^e_{\mathrm{p}}$ is the fluid pressure on the fracture surface, $\boldsymbol{F}^e_{\mathrm{s}}$ is the spring force, and $\boldsymbol{F}^e_{\mathrm{t}}$ is the force on the traction boundary.

The finite volume method is derived by obtaining the relationship between the flux Q_{ij} across the interface $\partial\Omega_{ij}$ and pressures p of the two adjacent cells. Let $K = k/\mu$ for simplicity. The flow velocities along line segments C_iC_0 and C_0C_j can be obtained from Darcy's law by Eqs. (2.6) and (2.7):

$$\boldsymbol{v}_{i0} = -K_i\nabla p_{i0} = -K_i\frac{p_0 - p_i}{D_i}(-\boldsymbol{d}_i) \tag{2.14a}$$

$$\boldsymbol{v}_{0j} = -K_j\nabla p = -K_{0j}\frac{p_j - p_0}{D_j}\boldsymbol{d}_j \tag{2.14b}$$

Flux across the interface $\partial\Omega_{ij}$ can be computed by taking the following integral:

$$Q_{ij} = Q_{i0} = \int \partial\Omega_{ij}\boldsymbol{v}_{i0} \cdot (-\boldsymbol{n}_i)dS = AK_i\frac{p_i - p_0}{D_i}(\boldsymbol{d}_i \cdot \boldsymbol{n}_i) \tag{2.15a}$$

$$Q_{ij} = Q_{0j} = \int \partial\Omega_{ij}\boldsymbol{v}_{0j} \cdot (\boldsymbol{n}_j)dS = AK_j\frac{p_0 - p_j}{D_j}(\boldsymbol{d}_j \cdot \boldsymbol{n}_j) \tag{2.15b}$$

where A is the interface area between adjacent elements, K_i is the inherent permeability of element i, D_i is the distance from the center of the element to the midpoint of the interface, $\boldsymbol{n}_i$ is the element normal vector of the interface, and $\boldsymbol{d}_i$ is the element direction vector along C_0C_i.

As the fracture criteria, the dual bilinear cohesive zone model foe fluid-driven propagation of multiscale tensile and shear fractures in tight reservoir is introduced (Wang & Zhang, 2022); in order to improve the computational efficiency, a multi-thread parallel computation method is also used (Wang et al., 2022).

2.4 Fractal Characterization Method for Fracture Network Morphology

The fracture network morphology formed by fluid-driven fracture propagation is affected by the distribution form of heterogeneous structures in rock mass; the quantitative description of the morphology of fracture network can help to better understand the relationship between the distribution of granules and natural fractures and fracture network. Many parameters can be used to describe the morphology of fractures, such as fracture occurrence, porosity, surface roughness, connectivity and so on; however, these indicators are difficult to use conventional geometrical features. Therefore, fractal dimension, as a quantitative index often used to describe the complexity and irregularity of the research object, has been applied to the geometrical characteristics analysis of the fracture network, focusing on the complexity of the fracture network (Xie, 1993). The fractal dimension of fracture network is computed by the box-counting method (Charkaluk et al., 1998; Xie, 1993), and the illustration of the fracturing fracture network and box-counting method for determining the fractal dimensions is shown in Fig. 2.7: firstly, the fracture result image is processed into a binary two-dimensional graph, and then the fracture binary image is covered with a grid with a side length of δ_k ($k = 1, 2,\ldots, n$); at the same time, the number of covered grids N_{δ_k} is counted, and the side length is halved to subdivide the grid into smaller grids δ_{k+1}; repetition of above process will derive a N_{δ_k} array corresponding to δ_k; Then the numbers of covering grids N_{δ_k} and δ_k may be drawn in the double logarithmic coordinate system, and the ln $N_{\delta_k} \sim (-\ln\ \delta_k)$ curve is generated; the slope of the straight line represents the fractal dimension D_B, which can realize the quantitative characterization of the fracture network as follows:

$$D_B = \lim_{k\to\infty} \frac{\ln\ N_{\delta_k}}{-\ln\ \delta_k} \tag{2.16}$$

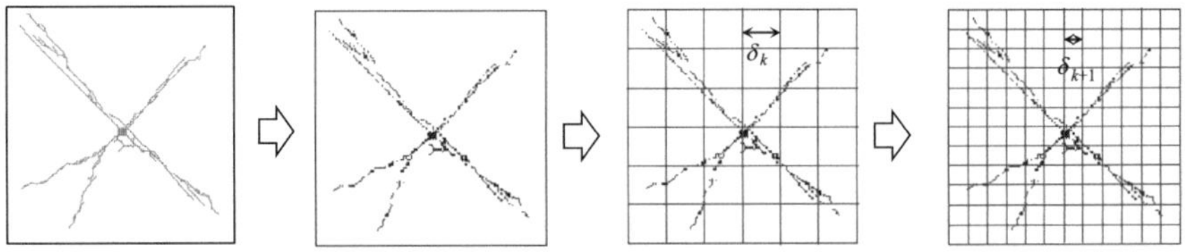

(a) Computed fracturing fracture network

(b) Binarized image of fracturing fracture network

(c) The kth covering with a grid size of δ_k

(d) The $(k+1)$th covering with a grid size of δ_{k+1}

Fig. 2.7 Illustration of the fracturing fracture network and box-counting method for determining the fractal dimensions

2.5 Global Procedure for Statistical Modelling, Fracture Propagation, and Fractal Characterization

The global procedure for dynamic intersection and deflection behaviours of hydraulic fractures meeting granules and natural fractures in tight reservoir rock based on statistical modelling and fractal characterization is shown in Fig. 2.8. Firstly, the combined finite element-discrete element-finite volume method is introduced and used for hydrofracturing in heterogeneous reservoir rock; using the statistical uniform distribution, the locations of granule and natural fracture models and cases under uniform distribution are established; Using the statistical Weibull distribution, the sizes (granule diameter or natural fracture length) of granule and natural fracture models and cases under Weibull distribution with variable λ_0 are established; subsequently, the fractal characterization method is introduced and used to quantitatively

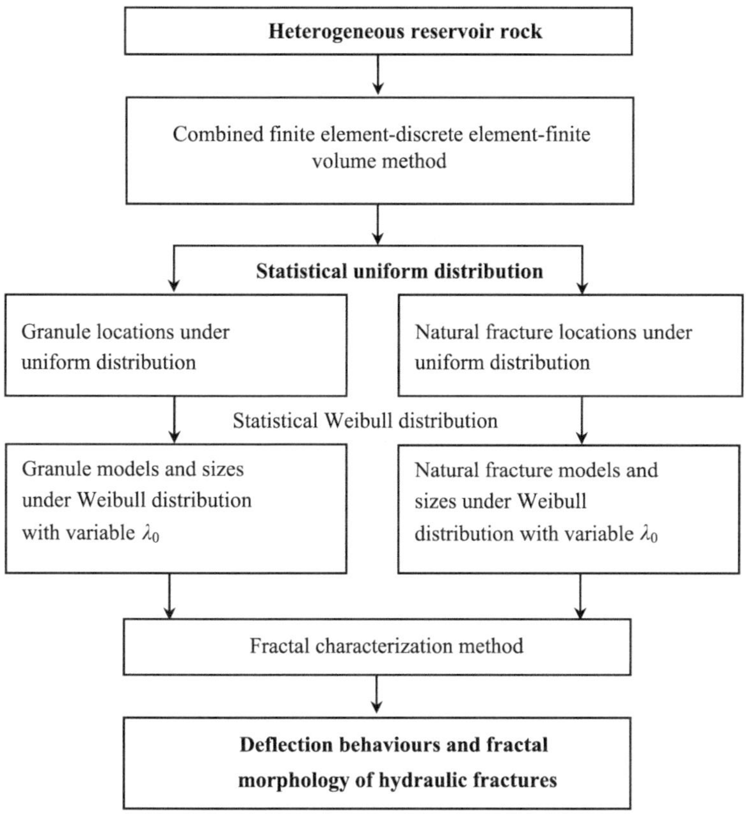

Fig. 2.8 Global procedure for dynamic intersection and deflection behaviours of hydraulic fractures meeting granules and natural fractures in tight reservoir rock based on statistical modelling and fractal characterization

evaluate the morphology of the computed fracture networks; finally, the deflection behaviours and fractal morphology of hydraulic fractures can be obtained, and the influences of granules and natural fractures on the intersection and deflection behaviours and fractal morphology of hydraulic fractures are also studied.

2.6 Results and Discussions

2.6.1 Propagation Behaviours and Fractal Characterization of Fracturing Fracture Network in Homogeneous Tight Reservoirs

To investigate the influences of heterogeneous granules and natural fractures on fluid driven-fracture propagation, a homogeneous model without granules and natural fractures is established for comparative analysis. The results of morphology of fracture network and pore pressure of homogeneous tight reservoirs at different time steps are shown in Table 2.5. High pore pressure areas appear in the reservoir due to fracturing fluid injection around the hydraulic fractures; due to the same in-situ stress in both directions (horizontal and vertical in-situ stresses in x and y directions are 60 MPa), the hydraulic fractures propagate almost simultaneously in both directions, and there is no long fracture in one direction; without the influence of heterogeneous factors, the fractures almost propagate straight. The results of morphology and fractal dimension of fracture network of homogeneous tight reservoirs are shown in Fig. 2.9. Figure 2.9a shows the final morphology of fracture network; based on this figure, the fractal dimension is calculated to obtain the curve shown in Fig. 2.9b, and the absolute value of the slope of the curve is the fractal dimension value of $D_B = 1.17$.

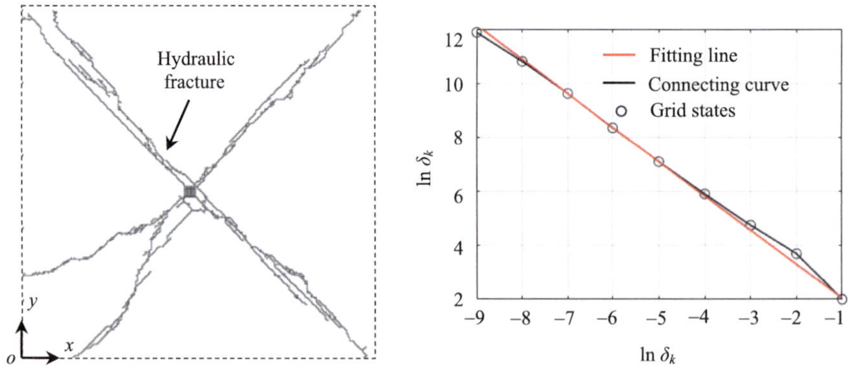

(a) Morphology of fracture network

(b) Homogeneous: D_B=1.17

Fig. 2.9 Results of morphology and fractal dimension of fracture network of homogeneous tight reservoirs

Table 2.5 Results of morphology of fracture network and pore pressure (Unit: MPa) of homogeneous tight reservoirs at different time steps *n*

n	Morphology of fracture network	Pore pressure
20000	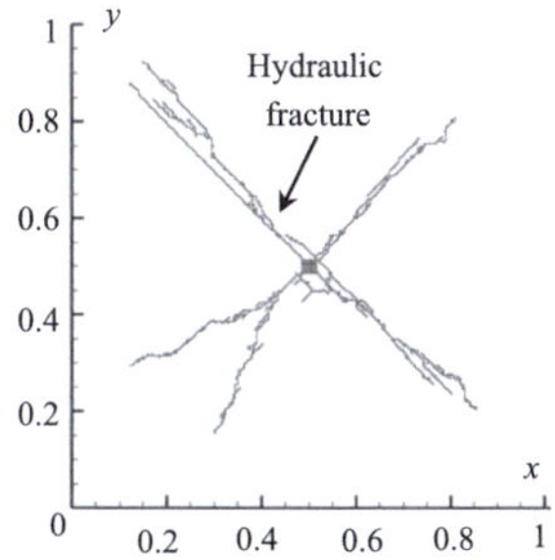	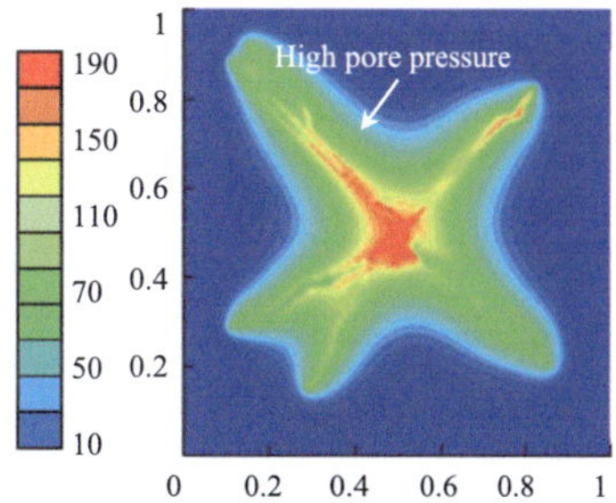
40000	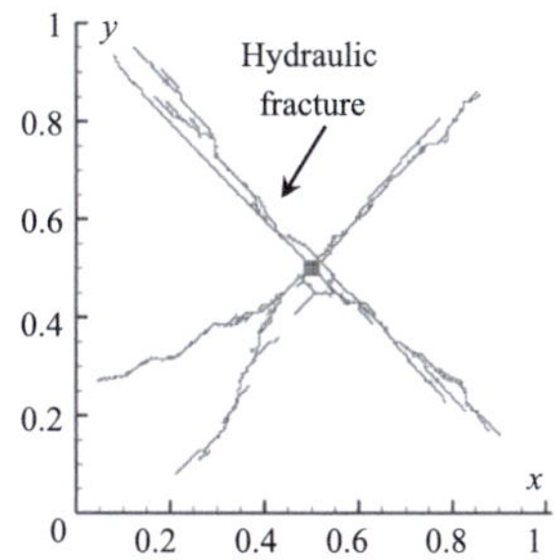	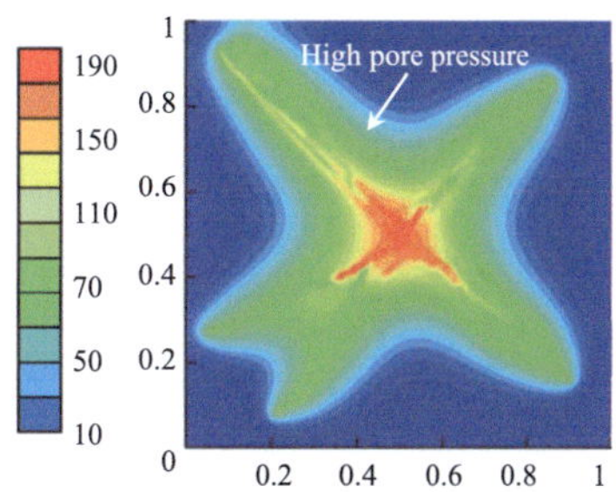
60000	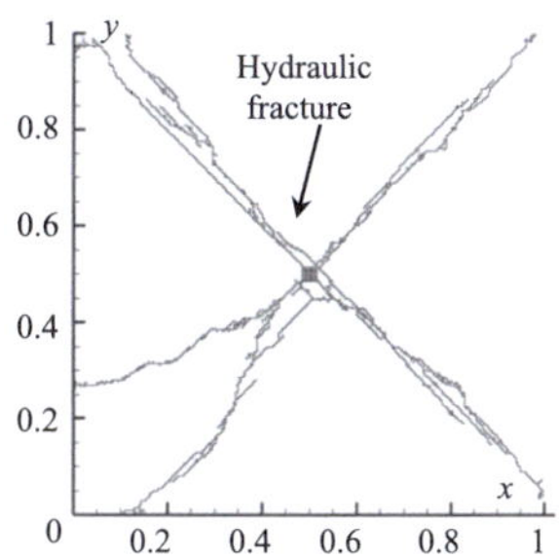	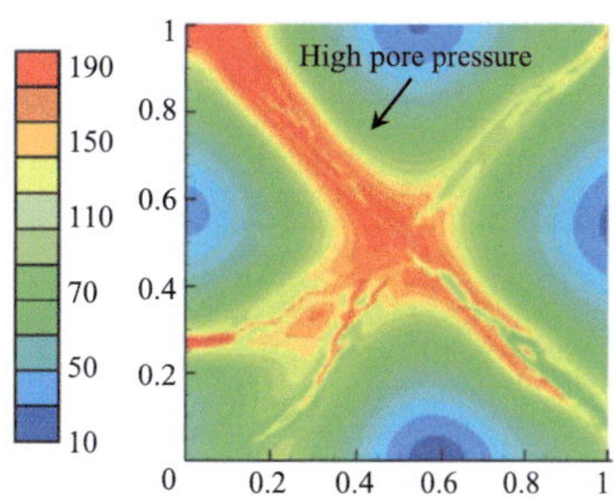

2.6.2 *Intersection and Deflection Behaviours of Hydraulic Fractures Meeting Granules*

In this section, the intersection and deflection behaviours of hydraulic fractures meeting granules are studied, and the numerical results of morphology of fracture network and pore pressure of heterogeneous tight reservoirs (Granule-I), and experimental results of fracture penetration, diversion, and arrest, are shown in Table 2.6. The high pore pressure areas also appear in the reservoir due to fracturing fluid injection around the hydraulic fractures; due to the same in-situ stress in both directions, the hydraulic fractures propagate almost simultaneously in both directions, however, there are some short and curved fractures arrested by the pre-existing granules. The intersection and deflection behaviours of hydraulic fractures when encountering granules in three different granule distribution models are shown in Fig. 2.10.

Table 2.6 Numerical results of morphology of fracture network and pore pressure of heterogeneous tight reservoirs (Granule-I), and experimental results of fracture penetration, diversion, and arrest

Numerical results	Experimental results
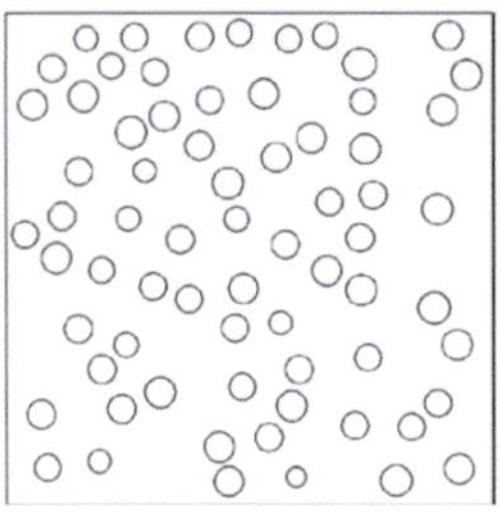 (a) Heterogeneous granules distribution	 (d) Fracture penetration
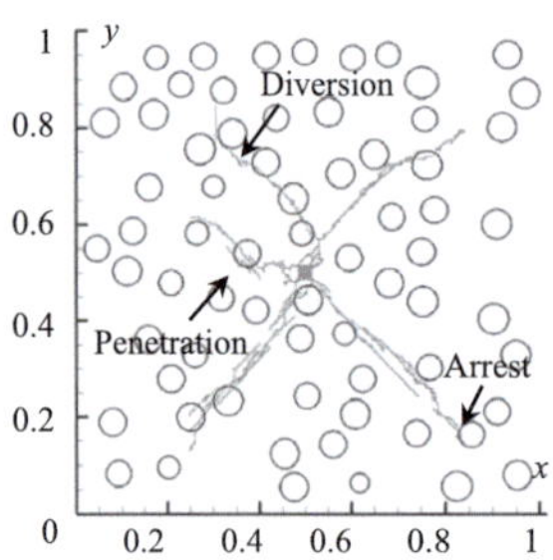 (b) Morphology of fracture network	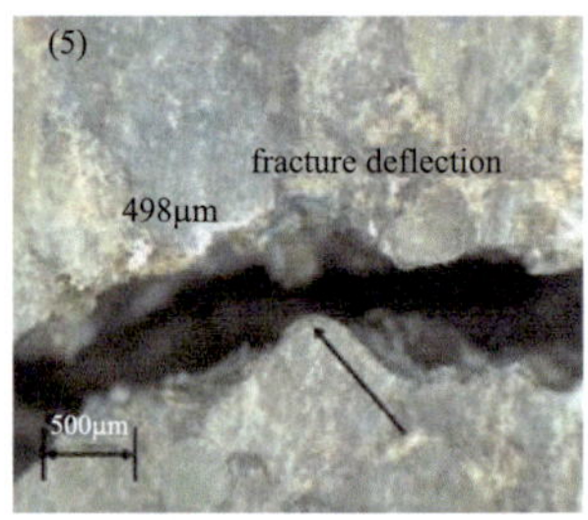 (e) Fracture deflection (diversion)
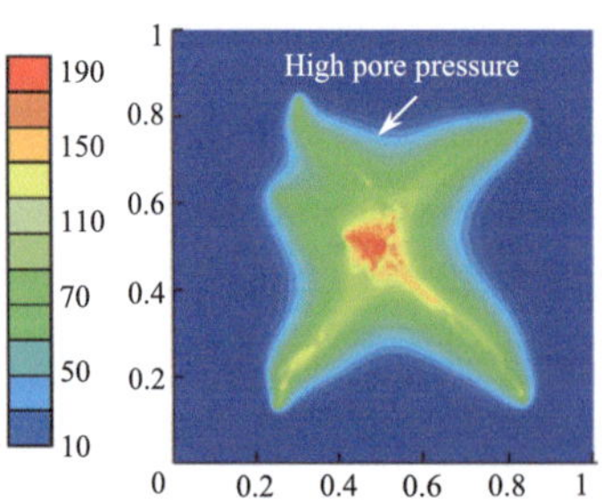 (c) Pore pressure (MPa)	 (f) Fracture arrest

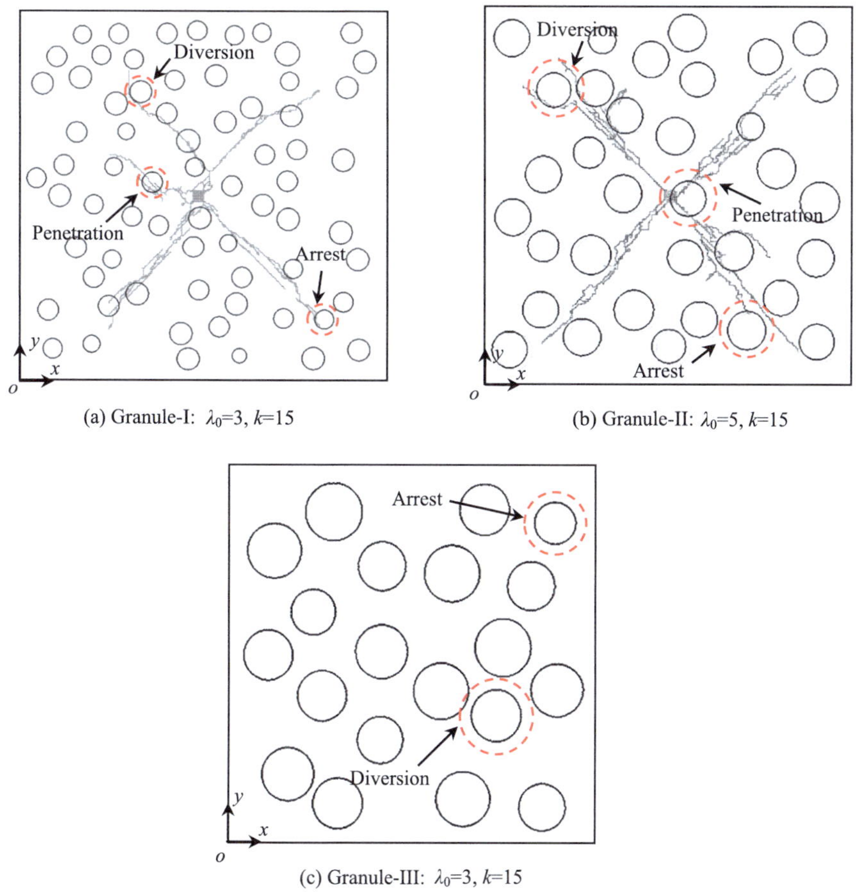

Fig. 2.10 The intersection and deviation behaviour of hydraulic fractures when encountering granules in three different granule distribution models

The following contents will explain and discuss the behaviours and quantitative fractal dimensions of these fracturing fracture networks:

(a) **Morphology of fracture network of hydraulic fractures meeting granules:** Due to the existence of granules on the right side of the fluid injection area, the hydraulic fractures initiate and propagate unsymmetrically, which indicates that the granules cause the direction of fracture. Once fracture begins to propagate, it easily penetrates the granules near the fluid injection area, however, then it may bypass the granule or be arrested, which indicates that the gravel penetration behaviour often occurs around the fluid injection area induced by the high pressure of fracturing fluid, while the gravel diversion and arrest behaviours are prone to occur in the areas where the fracturing fluid pressure decreases. In Figs. 2.10b and c, due to the large sizes of granules, there are large gaps and decentralized distribution between the granules; hydraulic fractures usually propagate in these gaps, and the statistical possibility of hydraulic fractures meeting granules is reduced; according to the large size of granule, it may take a lot of energy for fracture diversion, and has a strong shielding and blocking effects on the fracture propagation to form simple geometrical morphology of fracture network.

(b) **Intersection and deflection behaviours of hydraulic fractures meeting granules:** The hydraulic fracture begins to generate branches near the fluid injection area. Once the hydraulic fracture meets the obstructions composed of granules in the propagation process, the fracture will arise intersection and deflection behaviours of three types: penetration, diversion, and arrest. The occurrence of these types of behaviours is related to some controlling factors, such as the fluid pressure at the fracture tip, the fracture toughness of the rock mass, and the intersection angle between the fracture and granule. Considering the positive correlation factors, i.e., larger fluid pressure around the fracture tips, the lower the fracture toughness of the rock mass, and the intersection angle between the fracture and granule is closer to 90°, the hydraulic fractures are more likely to penetrate the granules; otherwise, the hydraulic fractures are prone to arise diversion behaviours and may even be arrested by granules. These intersection and deflection behaviours of hydraulic fractures meeting granules have also been monitored in physical experiments, as shown in Figs. 2.10d–f.

(c) **Fractal dimensions of morphology of fracture network of hydraulic fractures meeting granules:** The fractal dimensions of fracturing fracture network of heterogeneous tight reservoirs embedded granules are shown in Fig. 2.11, which are calculated to obtain the curves based on the figures of morphology of fracture network as shown in Fig. 2.10, and the absolute values of the slope of each curve are the fractal dimension values of $D_B = 1.26$, 1.22, and 1.19, respectively. The comparison of fractal dimensions of fracturing fracture network between homogeneous and heterogeneous tight reservoirs embedded granules is shown in Fig. 2.12. The fractal dimensions of the fracture network of the heterogeneous tight reservoirs (Granule-I, Granule-II, and Granule-III) are higher than that of the homogeneous model of $D_B = 1.17$, indicating that the existence of granules increases the complexity of hydrofracturing fracture network. The fractal dimension of fracture network in Granule-II is 1.22, which is smaller than that in Granule-I of $D_B = 1.26$; furthermore, the fractal dimension of fracture network in Granule-III is 1.19, which is smaller than that in Granule-II of $D_B = 1.22$. The complexity of fracture network decreases with the increase of the statistical average granule size, and the small granules are more likely to induce the penetration, diversion, and arrest behaviours of hydraulic fractures at tiny scale.

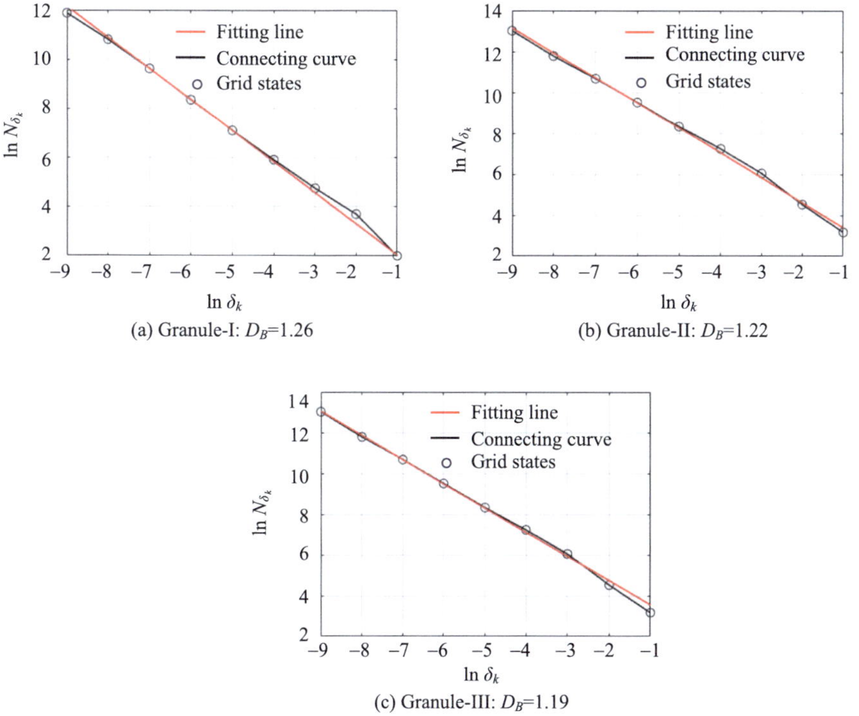

Fig. 2.11 Fractal dimensions of fracturing fracture network of heterogeneous tight reservoirs embedded granules

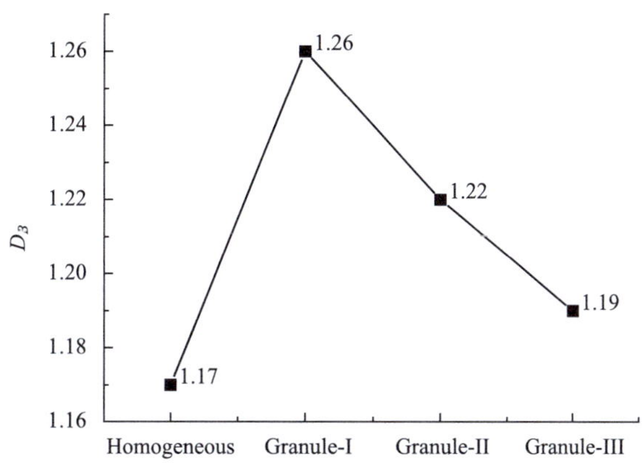

Fig. 2.12 Comparison of fractal dimensions of fracturing fracture network between homogeneous and heterogeneous tight reservoirs embedded granules

2.6.3 Intersection and Deflection Behaviours of Hydraulic Fractures Meeting Natural Fractures

In this section, the intersection and deflection behaviours of hydraulic fractures meeting natural fractures are studied, and the numerical results of morphology of fracture network and pore pressure of heterogeneous tight reservoirs (Fracture-I), and experimental results of Fracture penetration, diversion, and arrest, are shown in Table 2.7. Similar to the influence of granules, the high pore pressure areas also

Table 2.7 Numerical results of morphology of fracture network and pore pressure of heterogeneous tight reservoirs (Fracture-I), and experimental results of fracture penetration, diversion, and arrest

Numerical results	Experimental results
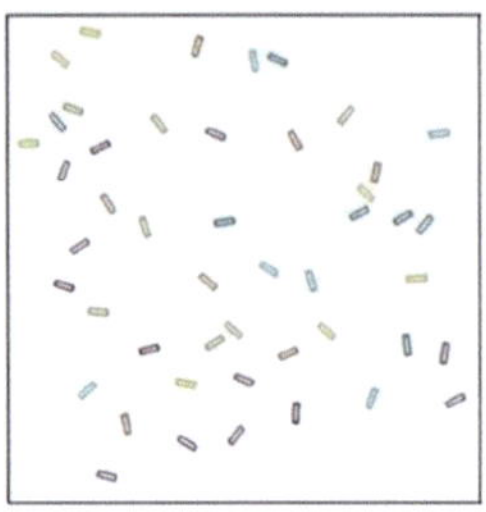 (a) Heterogeneous granules distribution	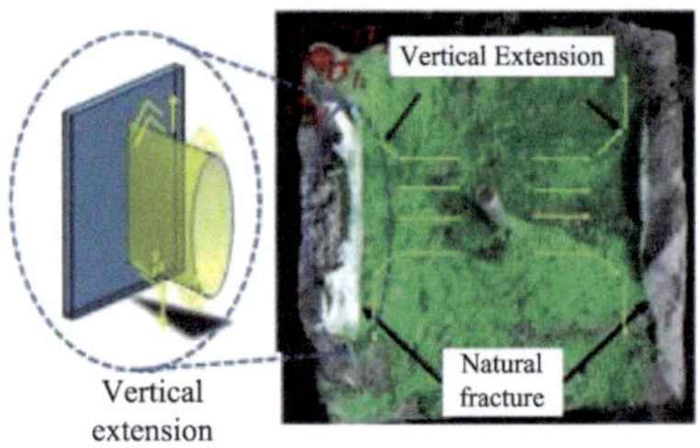 (d) Fracture propagation (penetration) along natural fracture
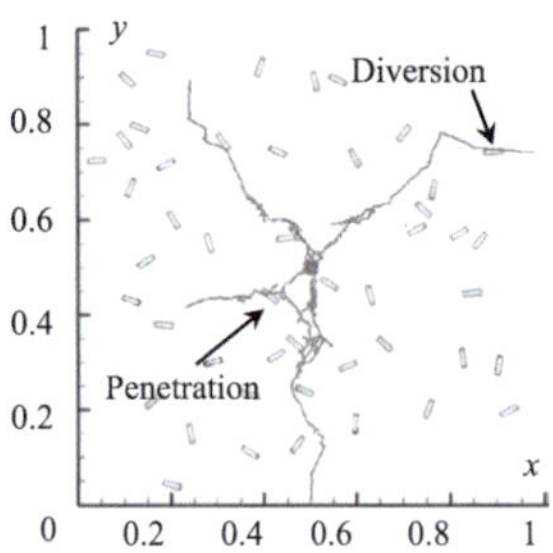 (b) Morphology of fracture network	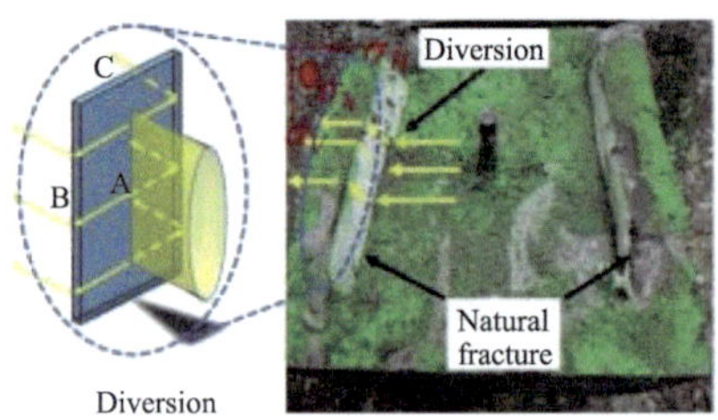 (e) Fracture diversion
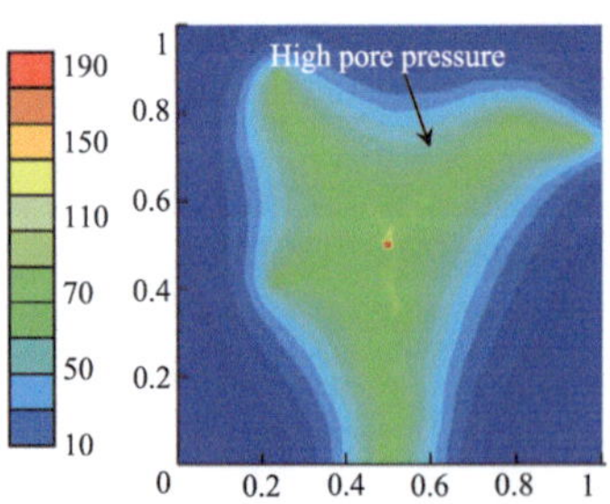 (c) Pore pressure(MPa)	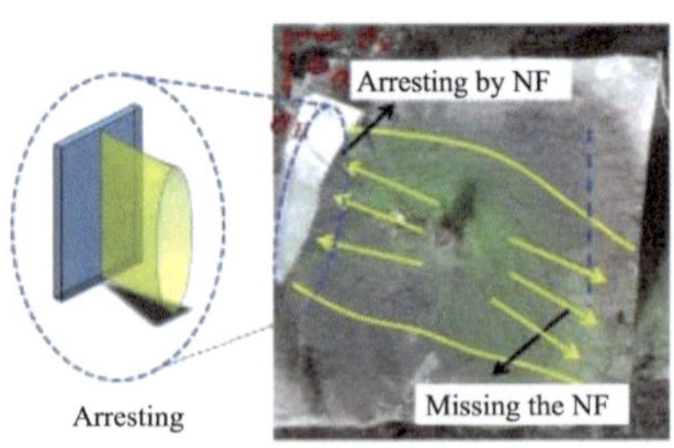 (f) Fracture arrest

appear in the reservoir due to fracturing fluid injection around the hydraulic fractures; there are some short and curved fractures arrested by the pre-existing natural fractures. The intersection and deflection behaviours of hydraulic fractures when encountering natural fractures in three different natural fractures distribution models are shown in Fig. 2.13.

The following contents will explain and discuss the behaviours and quantitative fractal dimensions of these fracturing fracture networks:

(a) **Morphology of fracture network of hydraulic fractures meeting natural fractures:** Due to the existence of natural fractures, the hydraulic fractures initiate and propagate unsymmetrically, which indicates that the natural fractures cause the direction of hydraulic fracture. Once fracture begins to propagate, it easily penetrates

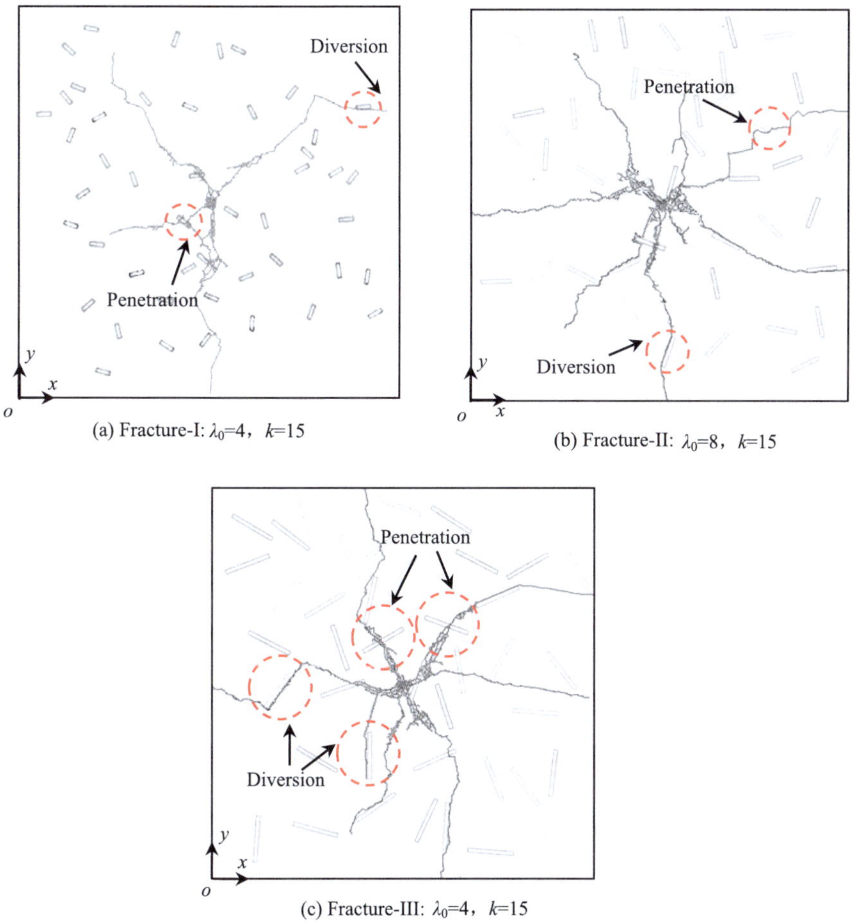

Fig. 2.13 The intersection and deviation behaviour of hydraulic fractures when encountering granules in three different natural fractures distribution models

the hydraulic fractures near the fluid injection area, however, then it may bypass the hydraulic fractures or propagate along the natural fractures, which indicates that the natural fractures penetration behaviour often occurs around the fluid injection area induced by the high pressure of fracturing fluid, while the gravel diversion and arrest behaviours are prone to occur in the areas where the fracturing fluid pressure decreases. In Figs. 2.13b and c, due to the large sizes of natural fractures, there are large gaps and decentralized distribution between the natural fractures; hydraulic fractures usually propagate in these gaps, and the statistical possibility of hydraulic fractures meeting natural fractures is reduced; according to the large size of natural fractures, it may take a lot of energy for fracture diversion, and has a strong shielding and blocking effects on the fracture propagation to form simple geometrical morphology of fracture network.

(b) **Intersection and deflection behaviours of hydraulic fractures meeting natural fractures:** The hydraulic fracture begins to generate branches near the fluid injection area. Once the hydraulic fracture meets the obstructions composed of natural fractures in the propagation process, the fracture will arise intersection and deflection behaviours of two types: penetration and diversion. When the angle between the propagation direction of hydraulic fractures and natural fractures is small, the hydraulic fractures may propagate along the direction of natural fractures and arise the diversion behaviours; once the angle between hydraulic fractures and natural fractures is close to 90°, the hydraulic fractures are prone to penetrate the natural fractures. Compared with the propagation behaviours of granule model, under the influence of natural fractures, the intersection and deflection of main fractures intensely vary; hydraulic fractures are easier to propagate along natural fractures and are less likely to be arrested by natural fractures, resulting in the formation of multiple longer fractures. The influences of granules and natural fractures on the propagation of hydraulic fractures in reservoir rock mass vary and demonstrate significant differences: Granules have a more inhibitory effect on hydraulic fractures, affecting their propagation; hydraulic fractures are easily connected to natural fractures, forming longer fractures, and promoting their propagation. These intersection and deflection behaviours of hydraulic fractures meeting natural fractures have also been monitored in physical experiments, as shown in Figs. 2.11d–f. It should be emphasized that natural fractures were found to prevent the propagation and extension of hydraulic fractures in the experiment, and the arrest behaviour occurred; however, this behaviour was not demonstrated in the simulation of this study, which is because the strength attribute of natural fractures set in this study is much weaker than that of rock matrix, and natural fractures are prone to be penetrated by hydraulic fractures.

(c) **Fractal dimensions of morphology of fracture network of hydraulic fractures meeting natural fractures:** The fractal dimensions of fracturing fracture network of heterogeneous tight reservoirs embedded natural fractures are shown in Fig. 2.14, which are calculated to obtain the curves based on the figures of morphology of fracture network as shown in Fig. 2.13, and the absolute values of the slope of each curve are the fractal dimension values of $D_B = 1.19, 1.21,$ and 1.22, respectively. The comparison of fractal dimensions of fracturing fracture network

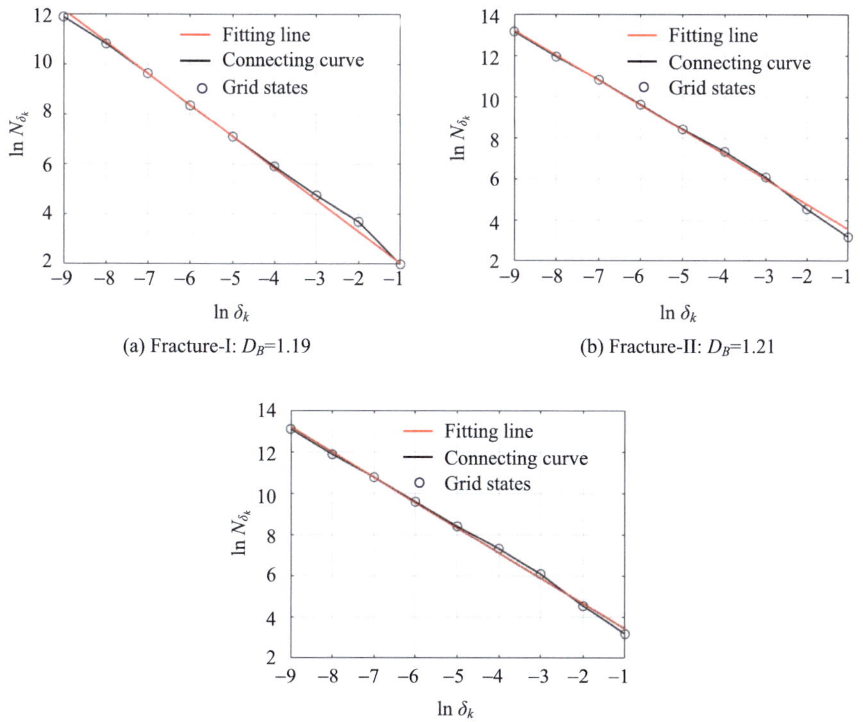

(a) Fracture-I: D_B=1.19

(b) Fracture-II: D_B=1.21

(c) Fracture-III: D_B=1.22

Fig. 2.14 Fractal dimensions of fracturing fracture network of heterogeneous tight reservoirs embedded natural fractures

between homogeneous and heterogeneous tight reservoirs embedded natural fractures is shown in Fig. 2.15. The fractal dimensions of the fracture network of the heterogeneous tight reservoirs (Fracture-I, Fracture-II, and Fracture-III) are higher than that of the homogeneous model of $D_B = 1.17$, indicating that the existence of natural fractures increases the complexity of hydrofracturing fracture network. The fractal dimension of fracture network in Fracture-II is 1.21, which is larger than that in Fracture-I of $D_B = 1.19$; furthermore, the fractal dimension of fracture network in Fracture-III is 1.22, which is larger than that in Fracture-II of $D_B = 1.21$. The complexity of fracture network increases with the increase of the statistical average size of natural fractures, and the larger natural fractures are more likely to induce the penetration and diversion behaviours of hydraulic fractures at large scale.

Based on the above analysis, a scheme can be proposed in the next future research to evaluate the complexity of hydrofracturing fracture networks by analyzing the statistical heterogeneous granules or natural fractures situation of reservoir rock. Firstly, multiple samples of heterogeneous reservoir rock need to be obtained, and geometrical information such as distribution and sizes of heterogeneous granules or natural fractures are detected through physical means; then, the special form (i.e.,

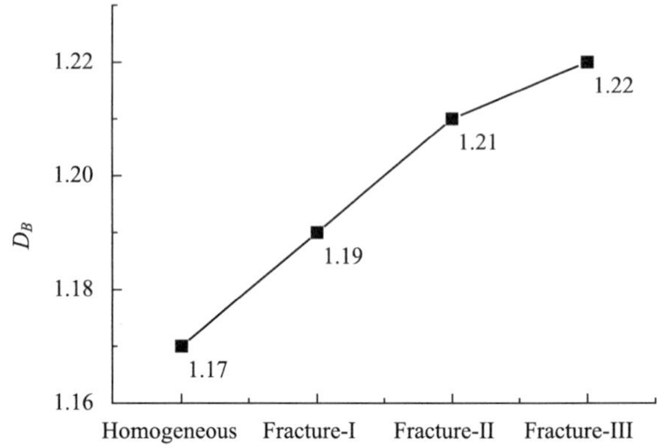

Fig. 2.15 Comparison of fractal dimensions of fracturing fracture network between homogeneous and heterogeneous tight reservoirs embedded natural fractures

average value λ_0 and concentration degree k in Weibull distribution) of statistical distribution may be analyzed, and the fractal dimension of the complex fracture network can be evaluated based on the statistical distribution.

2.7 Conclusions

The main conclusions of this study are as follows:

(1) To investigate the dynamic intersection and deflection behaviours of hydraulic fractures meeting granules and natural fractures in tight reservoir rock, the numerical models and cases based on statistical modelling and fractal characterization are proposed. The statistical modelling for tight heterogeneous reservoir rock, including statistical uniform and Weibull distribution of heterogeneous reservoir rock and establishment process of statistical models with granules and natural fractures are developed; the global procedure for statistical modelling, fracture propagation, and fractal characterization is established. Furthermore, using the combined finite element-discrete element-finite volume method, the dynamic intersection and deflection behaviours of hydraulic fractures meeting granules and natural fractures are investigated and analyzed.

(2) Once the hydraulic fracture meets the obstruction composed of granules in the propagation process, the fracture will arise intersection and deflection behaviours of three types: penetration, diversion, and arrest. Considering the positive correlation factors, i.e., larger fluid pressure around the fracture tips, the lower the fracture toughness of the rock mass, and the intersection angle between the fracture and granule is closer to 90°, the hydraulic fractures are more likely to penetrate the granules; otherwise, the hydraulic fractures are prone to arise diversion behaviours and may even be arrested by granules. The fractal dimension representing the complexity of a fracture network decreases with the increase of the statistical average granule size, and the small granules

are more likely to induce the penetration, diversion, and arrest behaviours of hydraulic fractures at tiny scale.

(3) Once the hydraulic fracture meets the obstructions composed of natural fractures in the propagation process, the fracture will arise intersection and deflection behaviours of two types: penetration and diversion. When the angle between the propagation direction of hydraulic fractures and natural fractures is small, the hydraulic fractures may propagate along the direction of natural fractures and arise the diversion behaviours; once the angle between hydraulic fractures and natural fractures is close to 90°, the hydraulic fractures are prone to penetrate the natural fractures. Granules have a more inhibitory effect on hydraulic fractures, affecting their propagation; hydraulic fractures are easily connected to natural fractures, forming longer fractures, and promoting their propagation. The fractal dimension representing complexity of fracture network increases with the increase of the statistical average size of natural fractures, and the larger natural fractures are more likely to induce the penetration and diversion behaviours of hydraulic fractures at large scale.

The intersection and deflection behaviours containing the penetration, diversion, and arrest in heterogeneous reservoirs rock are simulated and analysed. The investigation conclusions of fluid-driven fracture propagation with different geometrical distributions of granules and natural fractures can provide theoretical guidance for efficient and quantitative fractal dimension characterization of fracture propagation behaviours and control and optimization of fracturing effects in practical engineering in heterogeneous reservoirs. This study discusses the influences of the statistical geometrical distribution of granules and natural fractures on the propagation of hydraulic fractures, and the next stage of investigation will focus on more general heterogeneous properties, such as the geometrical form of beddings, geometrical multi-scale heterogeneity, and the geomechanical properties of heterogeneous rock masses on the propagation behaviours of hydraulic fractures.

References

Ali, S., Yan, C., Wang, T., Zheng, Y., Han, D., & Ke, W. (2024). Evaluating the impact of calcite and heterogeneity on the mechanical behavior of coal: A numerical study with grain-based finite-discrete element method. *Engineering Fracture Mechanics, 297*(21), Article 109880. https://doi.org/10.1016/j.engfracmech.2024.109880

Anderson, G. D. (1981). Effects of friction on hydraulic fracture growth near unbonded interfaces in rocks. *SPE Journal, 21*(1), 21–29. https://doi.org/10.2118/8347-PA

Berkowitz, B., & Hadad, A. (1997). Fractal and multifractal measures of natural and synthetic fracture networks. *Journal of Geophysical Research-Solid Earth, 102*(B6), 12205–12218. https://doi.org/10.1029/97JB00304

Biot, M. A. (1955). Theory of elasticity and consolidation for a porous anisotropic solid. *Journal of Applied Physics, 26*(2), 182–185. https://doi.org/10.1063/1.1721956

Blanton, T. L. (1982). An experimental study of interaction between hydraulically induced and pre-existing fractures. In *SPE Unconventional Resources Conference/Gas Technology Symposium*, SPE-10847, pp. 559–561. https://doi.org/10.2118/10847-MS.

Blanton, T. L. (1986). Propagation of hydraulically and dynamically induced fractures in naturally fractured reservoirs. In *SPE Unconventional Resources Conference/Gas Technology Symposium*, SPE-15261, pp. 613–627. https://doi.org/10.2118/15261-MS.

Boguna, M., Bonamassa, I., De Domenico, M., Havlin, S., Krioukov, D., & Serrano, M. Á. (2021). Network geometry. *Nature Reviews Physics, 3*(2), 114–135. https://doi.org/10.1038/s42254-020-00264-4

Bunger, A. P., Kear, J., Dyskin, A. V., & Pasternak, E. (2015). Sustained acoustic emissions following tensile crack propagation in a crystalline rock. *International Journal of Fracture, 193*, 87–98. https://doi.org/10.1007/s10704-015-0020-7

Cai, J., Wei, W., Hu, X., Liu, R., & Wang, J. (2017). Fractal characterization of dynamic fracture network extension in porous media. *Fractals, 25*(2), 1750023. https://doi.org/10.1142/S0218348X17500232

Charkaluk, E., Bigerelle, M., & Iost, A. (1998). Fractals and fracture. *Engineering Fracture Mechanics, 61*(1), 119–139. https://doi.org/10.1016/S0013-7944(98)00035-6

Daneshy, A. A. (1974). Hydraulic fracture propagation in the presence of planes of weakness. In *SPE Europec Featured at EAGE Conference and Exhibition*, SPE-4852, pp. 1–8. https://doi.org/10.2118/4852-MS.

Frosch, G. P., Tillich, J. E., Haselmeier, R., Holz, M., & Althaus, E. (2000). Probing the pore space of geothermal reservoir sandstones by Nuclear Magnetic Resonance. *Geothermics, 29*(6), 671–687. https://doi.org/10.1016/S0375-6505(00)00031-6

Geertsma, J., & De Klerk, F. (1969). A rapid method of predicting width and extent of hydraulically induced fractures. *Journal of Petroleum Technology, 21*(12), 1571–1581. https://doi.org/10.2118/2458-PA

Ghanbarian, B., & Hunt, A. G. (2017). Fractals: Concepts and Applications in Geosciences. CRC Press. https://doi.org/10.1201/9781315152264.

Gomaa, A. M., Zhang, B., Qu, Q., Nelson, S., & Chen, J. (2014). Using NMR technology to study the flow of fracture fluid inside shale formation. In *SPE International Symposium and Exhibition on Formation Damage Control*. https://doi.org/10.2118/168174-MS.

Hou, B., Zeng, C., Chen, D., Fan, M., & Chen, M. (2017). Prediction of wellbore stability in conglomerate formation using discrete element method. *Arabian Journal for Science and Engineering, 42*(4), 1609–1619. https://doi.org/10.1007/s13369-016-2346-5

Jiang, L., Liu, J., Liu, T., & Yang, D. (2020). Semi-analytical modeling of transient pressure behaviour for fractured horizontal wells in a tight formation with fractal-like discrete fracture network. *Geoenergy Science and Engineering, 197*, Article 107937. https://doi.org/10.1016/j.petrol.2020.107937

Khristianovich, S. A., & Zheltov, Y. P. (1955). Formation of vertical fractures by means of highly viscous liquid. In *World Petroleum Congress Proceedings*, pp. 579–586. https://onepetro.org/WPCONGRESS/proceedings-abstract/WPC04/All-WPC04/WPC-6132/203824.

Lan, H., Martin, C. D., & Hu, B. (2010). Effect of heterogeneity of brittle rock on micromechanical extensile behavior during compression loading. *Journal of Geophysical Research, 115*, B01202. https://doi.org/10.1029/2009jb006496

Li, L., Meng, Q., Wang, S., Li, G., & Tang, C. (2013). A numerical investigation of the hydraulic fracturing behaviour of conglomerate in Glutenite formation. *Acta Geotechnica, 8*(6), 597–618. https://doi.org/10.1007/s11440-013-0209-8

Li, M., Tang, S., Guo, T., & Qi, M. (2018). Numerical investigation of hydraulic fracture propagation in the glutenite reservoir. *Journal of Geophysics and Engineering, 15*(5), 2124–2138. https://doi.org/10.1088/1742-2140/aaba27

Li, M., Zuo, J., Hu, D., Shao, J., & Liu, D. (2020). Experimental and numerical investigation of microstructure effect on the mechanical behavior and failure process of brittle rocks. *Computers and Geotechnics, 125*, Article 103639. https://doi.org/10.1016/j.compgeo.2020.103639

Liu, K., & Ostadhassan, M. (2017). Quantification of the microstructures of Bakken shale reservoirs using multi-fractal and lacunarity analysis. *Journal of Natural Gas Science and Engineering, 39*, 62–71. https://doi.org/10.1016/j.jngse.2017.01.035

Liu, H. Y., Roquete, M., Kou, S. Q., & Lindqvist, P. A. (2004). Characterization of rock heterogeneity and numerical verification. *Engineering Geology, 72*(1–2), 89–119. https://doi.org/10.1016/j.enggeo.2003.06.004

Liu, H. Y., Kou, S. Q., Lindqvist, P. A., & Tang, C. A. (2006). Numerical modelling of the heterogeneous rock fracture process using various test techniques. *Rock Mechanics and Rock Engineering, 40*(2), 107–144. https://doi.org/10.1007/s00603-006-0091-x

Liu, Z., Wang, S., Zhao, H., Wang, L., Li, W., Geng, Y., Tao, S., Zhang, G., & Chen, M. (2018). Effect of random natural fractures on hydraulic fracture propagation geometry in fractured carbonate rocks. *Rock Mechanics and Rock Engineering, 51*, 491–511. https://doi.org/10.1007/s00603-017-1331-y

Ma, X. F., Zou, Y. S., & Li, N. (2017). Experimental study on the mechanism of hydraulic fracture growth in a glutenite reservoir. *Journal of Structural Geology, 97*, 37–47. https://doi.org/10.1016/j.jsg.2017.02.012

Manouchehrian, A., & Cai, M. (2016). Influence of material heterogeneity on failure intensity in unstable rock failure. *Computers and Geotechnics, 71*, 237–246. https://doi.org/10.1016/j.compgeo.2015.10.004

Mngadi, S. B., Durrheim, R. J., Manzi, M. S. D., Ogasawara, H., Yabe, Y., Yilmaz, H., Wechsler, N., Van Aswegen, G., Roberts, D., Ward, A. K., Naoi, M., Moriya, H., Nakatani, M., Ishida, A., Satreps Team & ICDP DSeis Team. (2019). Integration of underground mapping, petrology, and high-resolution microseismicity analysis to characterise weak geotechnical zones in deep South African gold mines. *International Journal of Rock Mechanics and Mining Sciences, 114*, 79–91.https://doi.org/10.1016/j.ijrmms.2018.10.003.

Mondal, D., Roy, P., & Behera, P. K. (2017). Use of correlation fractal dimension signatures for understanding the overlying strata dynamics in longwall coal mines. *International Journal of Rock Mechanics and Mining Sciences, 91*, 210–221. https://doi.org/10.1016/j.ijrmms.2016.11.019

Mousavi, N. M., Fisher, Q. J., Gironacci, E., & Rezania, M. (2018). Experimental study and numerical modeling of fracture propagation in shale rocks during Brazilian disk test. *Rock Mechanics and Rock Engineering, 51*(6), 1755–1775. https://doi.org/10.1007/s00603-018-1429-x

Movassagh, A., Haghighi, M., Zhang, X., Kasperczyk, D., & Sayyafzadeh, M. (2021). A fractal approach for surface roughness analysis of laboratory hydraulic fracture. *Journal of Natural Gas Science and Engineering, 85*, Article 103703. https://doi.org/10.1016/j.jngse.2020.103703

Olson, J. E., Bahorich, B., & Holder, J. (2012). Examining hydraulic fracture: Natural fracture interaction in hydrostone block experiments. In *SPE Hydraulic Fracturing Technology Conference*, SPE-152618. https://doi.org/10.2118/152618-MS.

Potyondy, D. O., & Cundall, P. A. (2004). A bonded-particle model for rock. *International Journal of Rock Mechanics and Mining Sciences, 41*(8), 1329–1364. https://doi.org/10.1016/j.ijrmms.2004.09.011

Rui, Z., Guo, T., Feng, Q., Qu, Z., Qi, N., & Gong, F. (2018). Influence of gravel on the propagation pattern of hydraulic fracture in the glutenite reservoir. *Geoenergy Science and Engineering, 165*, 627–639. https://doi.org/10.1016/j.petrol.2018.02.067

Sharafisafa, M., Aliabadian, Z., & Shen, L. (2020). Crack initiation and failure development in bimrocks using digital image correlation under dynamic load. *Theoretical and Applied Fracture Mechanics, 109*(5), Article 102688. https://doi.org/10.1016/j.tafmec.2020.102688

Wang, Y., & Liu, X. (2021). Stress-dependent unstable dynamic propagation of three-dimensional multiple hydraulic fractures with improved fracturing sequences in heterogeneous reservoirs: Numerical cases study via poroelastic effective medium model. *Energy & Fuels, 35*(22), 18543–18562. https://doi.org/10.1021/acs.energyfuels.1c03132

Wang, Y., & Zhang, X. (2022). Dual bilinear cohesive zone model-based fluid-driven propagation of multiscale tensile and shear fractures in tight reservoir. *Engineering Computations, 39*(10), 3416–3441. https://doi.org/10.1108/EC-01-2022-0013

Wang, T., Zhou, W., Chen, J., Xiao, X., Li, Y., & Zhao, X. (2014). Simulation of hydraulic fracturing using particle flow method and application in a coal mine. *International Journal of Coal Geology, 121*, 1–13. https://doi.org/10.1016/j.coal.2013.10.012

Wang, T., Hu, W., Elsworth, D., Zhou, W., Zhou, W., Zhao, X., & Zhao, L. (2017). The effect of natural fractures on hydraulic fracturing propagation in coal seams. *Geoenergy Science and Engineering, 150*, 180–190. https://doi.org/10.1016/j.petrol.2016.12.009

Wang, Y., Duan, Y., Liu, X., Huang, J., & Hao, N. (2021). Dynamic propagation and intersection of hydraulic fractures and pre-existing natural fractures involving the sensitivity factors: Orientation, spacing, length, and persistence. *Energy & Fuels, 35*(19), 15728–15741. https://doi.org/10.1021/acs.energyfuels.1c02896

Wang, Y., Wang, J., & Li, L. (2022). Dynamic propagation behaviors of hydraulic fracture networks considering hydro-mechanical coupling effects in tight oil and gas reservoirs: A multi-thread parallel computation method. *Computers and Geotechnics, 152*, Article 105016. https://doi.org/10.1016/j.compgeo.2022.105016

Warpinski, N. R. (1991). Hydraulic fracturing in tight, fissured media. *Journal of Petroleum Technology, 43*(2), 146–151. https://doi.org/10.2118/20154-PA

Warpinski, N. R., Mayerhofer, M. J., Vincent, M. C., Cipolla, C. L., & Lolon, E. P. (2009). Stimulating unconventional reservoirs: Maximizing network growth while optimizing fracture conductivity. *Journal of Canadian Petroleum Technology, 48*, 39–51. https://doi.org/10.2118/114173-PA

Wong, T. F., Wong, R. H. C., Chau, K. T., & Tang, C. A. (2006), Microcrack statistics, Weibull distribution and micromechanical modeling of compressive failure in rock. *Mechanics of Materials, 38*(7), 664–681.https://doi.org/10.1016/j.mechmat.2005.12.002.

Xie, H. (1993). Fractals in Rock Mechanics. CRC Press. https://doi.org/10.1201/9781003077626.

Xv, C., Zhang, G., Liu, Y., & Wang, P. (2019). Experimental study on hydraulic fracture propagation in conglomerate reservoirs. In *53rd U.S. Rock Mechanics/Geomechanics Symposium*, ARMA-2019-1844. https://onepetro.org/ARMAUSRMS/proceedings-abstract/ARMA19/All-ARMA19/ARMA-2019-1844/125012.

Yan, Y., Zhang, G., Li, S., & Nie, Y. (2019). Study on the influence of conglomerate meso-structures characteristics on crack propagation. In *53rd U.S. Rock Mechanics/Geomechanics Symposium*, ARMA-2019-1742. https://onepetro.org/ARMAUSRMS/proceedings-abstract/ARMA19/All-ARMA19/ARMA-2019-1742/124965

Yue, Z., Chen, S., & Tham, L. (2003). Finite element modeling of geomaterials using digital image processing. *Computers and Geotechnics, 30*(5), 375–397. https://doi.org/10.1016/S0266-352X(03)00015-6

Zhang, B., Liu, J. Y., Wang, S. G., Li, S. C., Yang, X. Y., Li, Y., Zhu, P. Y., & Yang, W. M. (2018a). Impact of the distance between pre-existing fracture and wellbore on hydraulic fracture propagation. *Journal of Natural Gas Science and Engineering, 57*, 155–165. https://doi.org/10.1016/j.jngse.2018.07.004

Zhang, Y., Zhang, J., Yuan, B., & Yin, S. (2018b). In-situ stresses controlling hydraulic fracture propagation and fracture breakdown pressure. *Geoenergy Science and Engineering, 164*, 164–173. https://doi.org/10.1016/j.petrol.2018.01.050

Zhao, Y., Wang, C., Ning, L., Zhao, H., & Bi, J. (2022). Pore and fracture development in coal under stress conditions based on nuclear magnetic resonance and fractal theory. *Fuel, 309*, Article 122112. https://doi.org/10.1016/j.fuel.2021.122112

Chapter 3
Deflection Behaviours and Fractal Morphology of Hydraulic Fractures Meeting Beddings and Granules with Variable Geometrical Configurations and Geomechanical Properties

3.1 Introduction

Hydrofracturing is the main technology for achieving efficient development of unconventional oil and gas resources (Chen et al., 2021; Song et al., 2017, 2020). The morphology of hydraulic fractures is mainly affected by deflection behaviours of fractures meeting embedded heterogeneous and discontinuous geological structures, such as the beddings and granules (Chen et al., 2021; Sun et al., 2021). Evaluating the propagation behaviours (deflection, intersections, penetration, and arrest) and morphology of hydraulic fractures is a key scientific issue to control and optimize the fracturing effects (Bower & Zyvoloski, 1997; Shuai et al., 2023; Su et al., 2015). It is urgent to quantitatively investigate the deflection behaviours and quantitative morphology of hydraulic fractures in heterogeneous rock mass and analyze the influences of heterogeneity on fracturing effects.

The bedding plane affects the dynamic propagation of fractures (Wang & Wang, 2021); the fracture toughness of shale is different under different bedding dip angles and loading directions, in laboratory experiments and numerical simulations, the fracture toughness parallel to the bedding direction is the lowest, and the fracture toughness perpendicular to the bedding direction is the highest, which indicates that the fracture toughness increases with the increase of the angle of the bedding (Meng et al., 2021). Once the fracture meets multiple parallel weak bedding planes, fracture arrest, penetration and deflection may occur, in which the mechanisms are complicated and the evaluation of hydraulic fracture networks has become challenging (Chen et al., 2018; Zhang et al., 2020). Due to the influence of granules in reservoirs, the fracture propagation mechanisms of granule and fine-grained rock are different, which exhibit completely different mechanical behaviors from conventional homogenized rock masses. Granules have an inhibitory effect on the propagation of fracturing fractures, hence, several propagation forms of hydraulic fractures when meeting granules are deflection, penetration, and arrest (Zhang et al., 2019). When the content of granule is more, the plasticity of the granule is stronger (Li et al.,

Y. Wang, *Deflection of Fracturing Fracture Network Disturbed by Discontinuity and Multiple Fractures in Rock*, https://doi.org/10.1007/978-981-95-1569-1_3

2018); the heterogeneity and plasticity index can describe the heterogeneity of rock micro-geometrical structure caused by the change of granules size in intact rock (Liu et al., 2018). The reduction of fracture pressure is more affected by granule heterogeneity; once the granule heterogeneity is large, the fracture initiation and propagation pressure are small; the initiation behaviours of hydraulic fractures is influenced by granule heterogeneity (Han et al., 2018). Through previous researches, it has been found that the geometrical configuration, density, and geomechanical properties of heterogeneity induced by beddings and granules in reservoir rock mass all affect the initiation, deflection, and propagation behaviours of hydraulic fractures, which are important factors that make it difficult to evaluate the fracture network. Previous studies mainly focused on the analysis of the influence of granule and bedding on fracture propagation, and few studies have used quantitative characterization to study the morphology of hydraulic fractures meeting beddings and granules, which is also a current challenge.

For investigating the deflection behaviours and fractal morphology of hydraulic fractures meeting beddings and granules, some experiments on hydrofracturing have also been carried out in the laboratory. For example, Zou et al. (2016) studied the influence of the bedding plane on the hydrofracturing fracture network and found that the bedding plane increased the complexity of fractures; in addition, to better study the propagation of the hydraulic fracture network in experiments, the acoustic emission, CT scanning, and fractal dimension are used to study the fracture morphology (Li et al., 2019; Liang et al., 2022; Zhou et al., 2018). Considering the complexity of hydrofracturing in practical heterogeneous reservoirs, it is difficult to obtain the whole process of fracture propagation and evolution by traditional field monitoring and laboratory experiments, therefore, numerical methods and models were introduced, and heterogeneous concrete models were used for macro-meso-micro three-scale failure simulation (Fu et al., 2013; McClure et al., 2016; Nguyen et al., 2012). The hydraulic coupling finite element-discrete element method, three-dimensional fully coupled propagated finite element method, high-order discontinuous Galerkin cohesive zone model were established for simulating the behaviours of hydraulic fractures (Gao & Ghassemi, 2020; Giovanardi et al., 2020; Lisjak et al., 2017). Furthermore, in this study, to investigate the deflection behavior and quantitative morphology of hydraulic fractures meeting bedding and granules, the numerical models and cases with different geometrical configurations and geomechanical properties will be proposed, by comprehensively using the combined finite element-discrete element-finite volume method and fractal characterization method.

The structure of this chapter is as follows: Sect. 3.2 introduces the governing partial differential equations and numerical discretization. Section 3.3 introduces the fractal morphology of fracturing fracture network based on fractal characterization method. Section 3.4 introduces the global procedure for deflection behaviours and fractal morphology of hydraulic fractures meeting beddings and granules. Section 3.5 introduces the numerical models and cases of beddings and granules with variable geometrical configurations and geomechanical properties. Section 3.6 introduces the results and discussion. Section 3.7 summarizes the conclusions of the study.

3.2 Governing Partial Differential Equations and Numerical Discretization

The following introductions involve the basic governing partial differential equations and numerical discretization, which were discussed in the previous chapter. For the completeness of the content, the equations are also provided in this section, but the detailed explanations of the parameters involved will not be repeated.

3.2.1 *Governing Equation of Solid Deformation*

Considering the influence of dynamic inertia, the mechanical equilibrium equation of solid deformation on representative elements in Cartesian coordinates is:

$$\nabla \cdot \boldsymbol{\sigma} = \rho \ddot{\boldsymbol{u}} + c\dot{\boldsymbol{u}} - \boldsymbol{f}, \quad x, y, z \in \Omega \tag{3.1}$$

3.2.2 *Governing Equations of Fluid Flow in Fractured Porous Media*

Assuming that the fluid in the fractured porous media is a single-phase flow, the simplified Darcy's law of fluid flow in porous media and hydraulic fractures can be used, and the velocity field can be obtained:

$$v_m = -\frac{k_m}{\mu}\nabla p_m, \quad x, y, z \in \Omega \tag{3.2}$$

$$v_f = -\frac{k_f}{\mu}\nabla p_f, \quad x, y, z \in \Omega \tag{3.3}$$

The pressure equation of single-phase flow in porous media and hydraulic fractures can be written as follows:

$$S_m \dot{p}_m - \nabla \cdot v_m = -\alpha \frac{\partial \varepsilon_V}{\partial t}, \quad x, y, z \in \Omega \tag{3.4}$$

$$S_f \dot{p}_f - \nabla \cdot v_f = q_f, \quad x, y, z \in \Omega \tag{3.5}$$

3.2.3 Numerical Discretization

Using the variational formula to solve the solid deformation control equation, Eq. (3.1) can be transformed into the following matrix form on the element e:

$$\boldsymbol{M}^e\ddot{\boldsymbol{D}}(t) + \boldsymbol{C}^e\dot{\boldsymbol{D}}(t) + \boldsymbol{K}^e\boldsymbol{D}(t) = \boldsymbol{F}^e, \quad x, y, z \in \Omega \tag{3.6}$$

As the fracture criteria, the dual bilinear cohesive zone model for fluid-driven propagation of multiscale tensile and shear fractures in tight reservoir is introduced (Wang & Zhang, 2022); in order to improve the computational efficiency, a multi-thread parallel computation method is also used (Wang et al., 2022).

3.3 Fractal Morphology of Fracturing Fracture Network Based on Fractal Characterization Method

Previous studies have shown that fractal dimension can well characterize the morphology of fracture networks (Xie, 2020). The fractal dimension of fracture network is computed by the box-counting method: firstly, the fracture result image is processed into a binary two-dimensional graph, and then the fracture binary image is covered with a grid with a side length of δ_k (k = 1, 2,…, n); at the same time, the number of covered grids N_{δ_k} is counted, and the side length is halved to subdivide the grid into smaller grids δ_{k+1}; repetition of above process will derive a N_{δ_k} array corresponding to δ_k, as shown in Fig. 3.1; Then the numbers of covering grids N_{δ_k} and δ_k may be drawn in the double logarithmic coordinate system, and the $\ln N_{\delta_k} \sim (-\ln \delta_k)$ curve is generated; the slope of the straight line represents the fractal dimension D_B, which can realize the quantitative characterization of the fracture network as follows:

$$D_B = \lim_{k\to\infty} \frac{\ln N_{\delta_k}}{-\ln \delta_k} \tag{3.7}$$

3.4 Global Procedure for Deflection Behaviours and Fractal Morphology of Hydraulic Fractures Meeting Beddings and Granules

The global procedure for deflection behaviours and fractal morphology of hydraulic fractures meeting beddings and granules with different geometrical configurations and geomechanical properties is shown in Fig. 3.2. Firstly, the combined finite

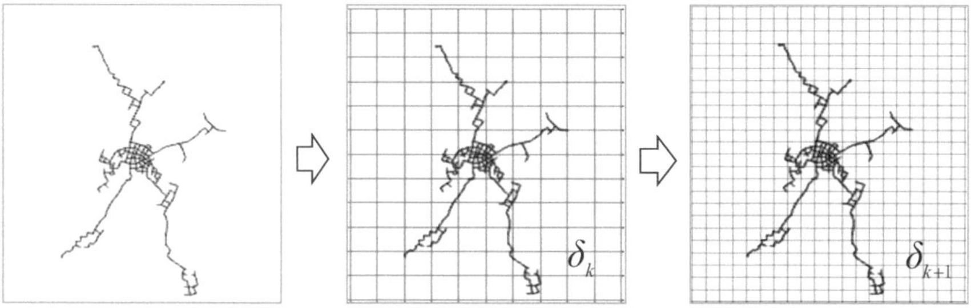

Fig. 3.1 Diagrammatic sketch of fractal characterization method using box-counting method and fractal morphology of fracture network

element-discrete element-finite volume method is introduced and used for hydrofracturing in heterogeneous reservoir rock; for variable geometrical configurations, the bedding models and cases with three different dip angles of $\beta = 0°, 45°, 90°$ and granule models and cases with three different configurations are established; for variable geomechanical properties, the bedding and granule models and cases with different geomechanical properties are established; subsequently, the fractal characterization method is introduced and used to quantitatively evaluate the morphology of the computed fracture networks; finally, the deflection behaviours and fractal morphology of hydraulic fractures can be obtained, and the influences of beddings and granules with different geometrical configurations and geomechanical properties on the deflection behaviours and fractal morphology of hydraulic fractures are also studied.

3.5 Numerical Models and Cases of Heterogeneous Reservoirs

3.5.1 Beddings with Variable Geometrical Configurations and Geomechanical Properties

To investigate the influence of bedding with variable geometrical configurations and geomechanical properties on the deflection behaviours and fractal morphology of hydraulic fractures, three geometrical models and finite element models with three different dip angles of β (0°, 45°, 90°) are established, and each model has 6 bedding planes with vertical spacing between the bedding planes is 13.43 m. The geometrical and finite element models for bedding with representative geometrical configuration of dip angle $\beta = 45°$ is shown in Fig. 3.3, and the other two types of models are similar and will not be elaborated here. The size of the model is 100 m × 100 m × 1 m, and the fracturing fluid injection perforation is located in the geometrical center of the model, and the eight-node hexahedral element is used in finite element

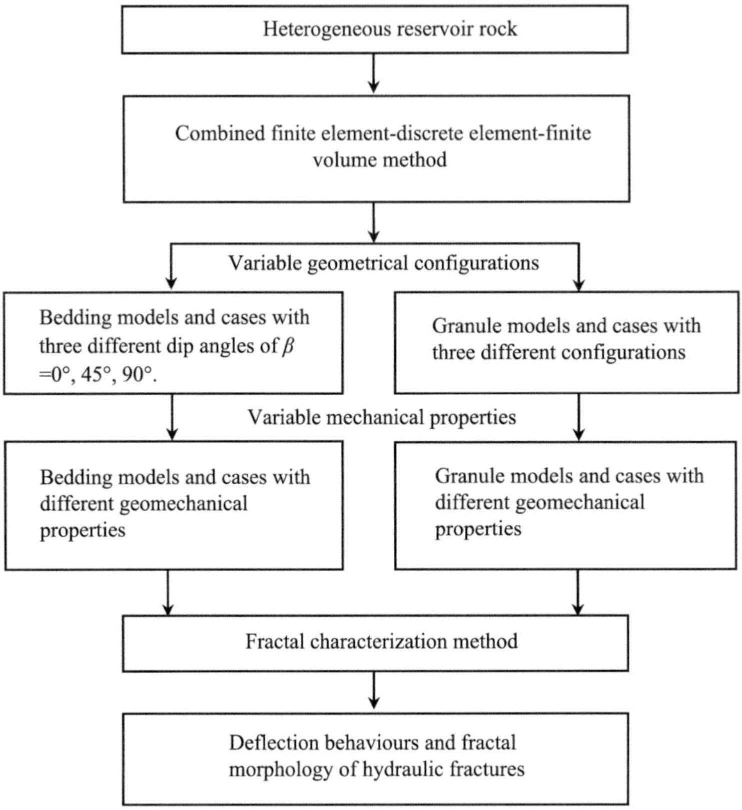

Fig. 3.2 Global procedure for deflection behaviours and fractal morphology of hydraulic fractures meeting beddings and granules

model. The in-situ stresses parallel to the x-axis and y-axis are 60 MPa. The basic geomechanical parameters of rock matrix and computation parameters are shown in Table 3.1, and the beddings and granules with variable geomechanical properties: Young's modulus, tensile strength, and cohesion are shown in Table 3.2, in which the variable geomechanical properties are used to detect the influence of heterogeneous properties on fracture propagation.

3.5.2 Granules with Variable Geometrical Configurations and Geomechanical Properties

The Weibull distribution is used to describe the random distribution of granule geometric parameters, and the probability density function is as follows:

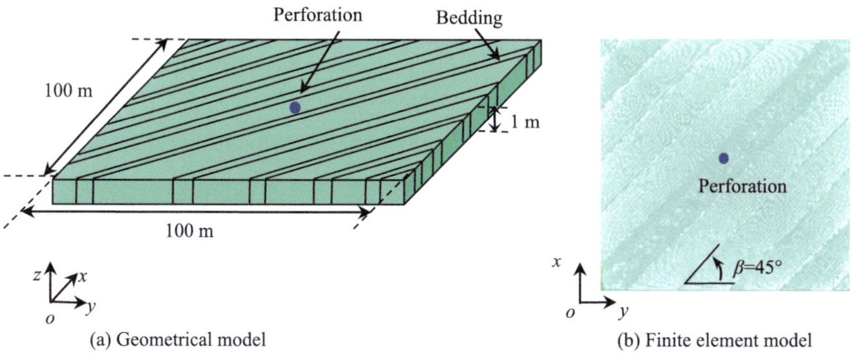

Fig. 3.3 Geometrical and finite element models for beddings with representative geometrical configuration of dip angle $\beta = 45°$

Table 3.1 Basic geomechanical parameters of rock matrix and computation parameters

Parameters	Value
Flow velocity $Q/(m^3/s)$	0.0001
Young's modulus of rock matrix E_m/GPa	22.38
Tensile strength of rock matrix σ_{mt}/MPa	8.25
Cohesive of rock matrix c_m/MPa	14.6
Poisson's ratio of rock matrix v	0.2
Internal friction angle of rock matrix $\varphi/(°)$	45
Density $\rho/(g/\ cm^3)$	2.5
Time of each loading step s/s	0.01
Loading step for bedding models n	10,000
Loading step for granule models n	50,000

Table 3.2 Beddings and granules with variable geomechanical properties: Young's modulus, tensile strength, and cohesion

Cases	Young's modulus E/GPa	Tensile strength σ_t/MPa	Cohesion c/MPa
I	4.5	1.7	2.9
II	22.38	8.25	14.6
III	112	41.25	73
IV	30	25	20
V	55	30	40
VI	85	35	60

$$f(\lambda; \lambda_0; k) = \begin{cases} 0 & \lambda < 0 \\ \frac{k}{\lambda_0}\left(\frac{\lambda}{\lambda_0}\right)^{k-1} e^{-(\lambda/\lambda_0)^k} & \lambda \geq 0 \end{cases} \tag{3.8}$$

where λ is the distribution parameter of heterogeneous components; λ_0 is the average value of distribution parameter; k is the concentration degree of distribution. To investigate the influence of granule with variable geometrical configurations and geomechanical properties on the deflection behaviours and fractal morphology of hydraulic fractures, three geometrical models and finite element models with different distribution and gradually increasing granule size (Configurations A, B, and C) are established based on the Weibull distribution parameters shown in Table 3.3, the models are shown in Figs. 3.4, 3.5 and 3.6, respectively. The in situ stresses parallel to the x-axis and y-axis are 10 MPa. The basic geomechanical parameters of rock matrix, computation parameters, and granules with variable geomechanical properties are shown in Tables 3.1 and 3.2, respectively.

Table 3.3 Weibull distribution parameters for variable geomechanical properties and configurations

Configurations	λ_0	k
A	3	15
B	5	15
C	7	15

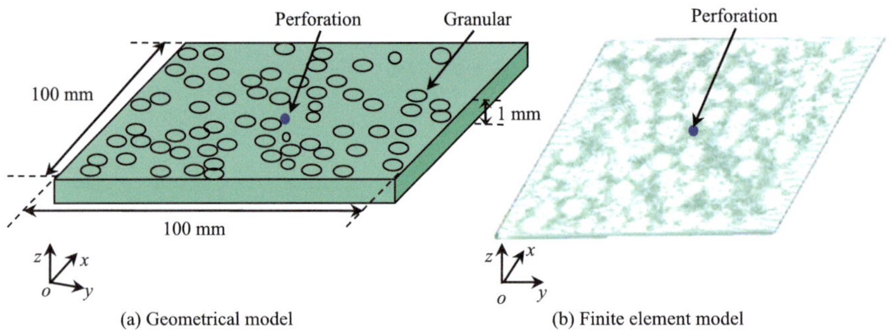

Fig. 3.4 Geometrical and finite element models for granules with geometrical configuration A

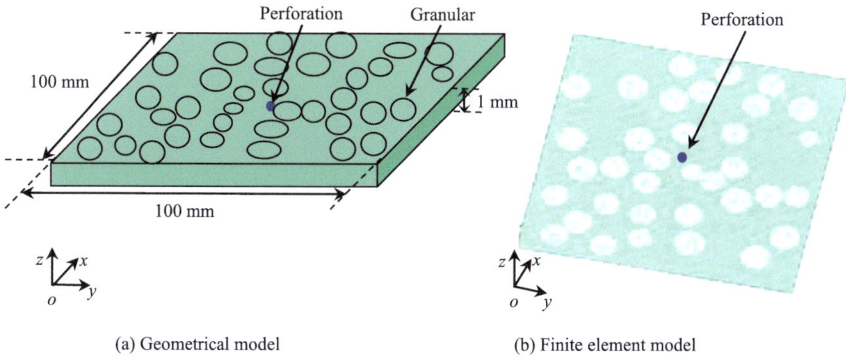

Fig. 3.5 Geometrical and finite element models for granules with geometrical configuration B

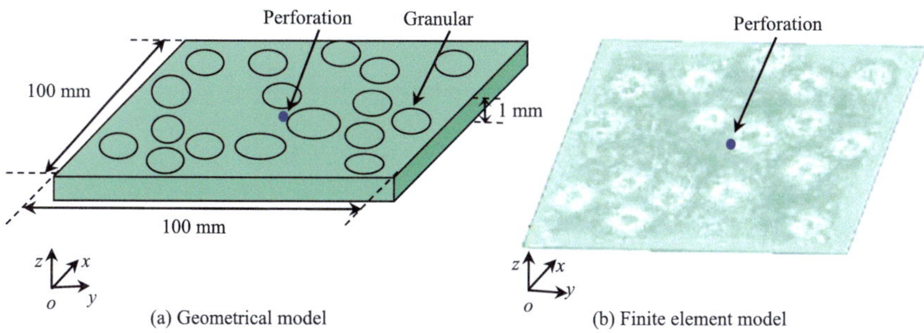

Fig. 3.6 Geometrical and finite element models for granules with geometrical configuration C

3.6 Results and Discussion

3.6.1 Beddings with Variable Geometrical Configurations

Figures 3.7, 3.8 and 3.9 show the results of deflection behaviours and morphology of hydraulic fractures meeting beddings with variable geometrical configurations in Cases I, II, and III, respectively. It can be seen that when $\beta = 0°$ (Fig. 3.7a) and 90° (Fig. 3.7c), the hydraulic fractures are more likely to penetrate the bedding plane, in which there are more fractures perpendicular to the bedding plane. The horizontal and vertical in-situ stresses in these models are the same, i.e., the ratio of in-situ stresses is 1:1, which makes the fracture network morphology similar for the cases of $\beta = 0°$ and 90°. At the case of $\beta = 45°$ (Fig. 3.7b), there are many fractures that deflected, which is because that when the fractures have a certain acute angle with the bedding plane, they are most likely to deviate along the bedding plane.

In Fig. 3.8, as the geomechanical properties (Young's modulus, tensile strength, and cohesion) of the bedding planes increase for improving their stiffness and strength, it can be seen that the fractures are prone to be influenced by the bedding planes, resulting in many deflections and branches; in Fig. 3.9, as the geomechanical

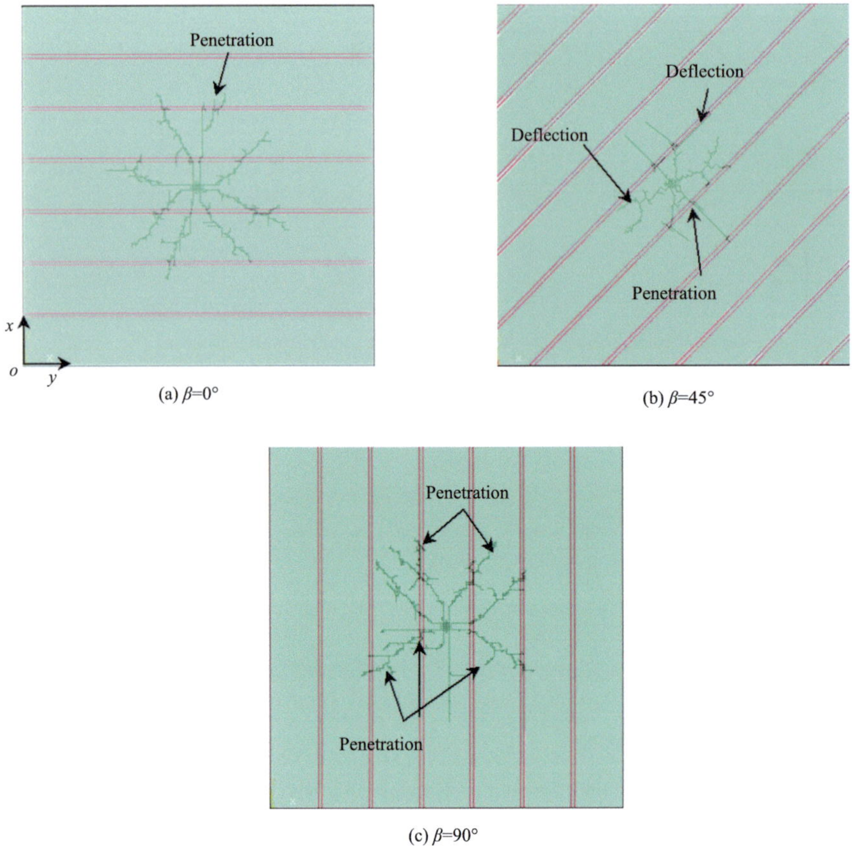

Fig. 3.7 Deflection behaviours and morphology of hydraulic fractures meeting beddings with variable geometrical configurations in Case I

properties of the bedding planes continue to increase, it can be seen that the fractures are further affected by the bedding planes, resulting in more deflection, and the fractures are hindered from propagating when they encounter the bedding planes. In Figs. 3.8 and 3.9, penetration and arrest are prone to occur at $\beta = 0°$ (Figs. 3.8a and 3.9a) and 90° (Figs. 3.8c and 3.9c), and fracture deflection is prone to occur at $\beta =$ 45° (Figs. 3.8b and 3.9b).

Figure 3.10 shows the fractal dimensions for morphology of hydraulic fractures meeting beddings with variable geometrical configurations, in which the following conclusions can be drawn:

(a) The results for the cases of $\beta = 0°$ and 90° are relatively close, because the in-situ stresses mentioned above are equal in both horizontal and vertical coordinate directions, and the fractal dimensions of the hydraulic fracture propagation morphology at the both bedding angles are similar; however, due to factors such as the randomness of hydraulic fracture propagation and different mesh generation in horizontal and vertical coordinate directions, the fractal dimensions of

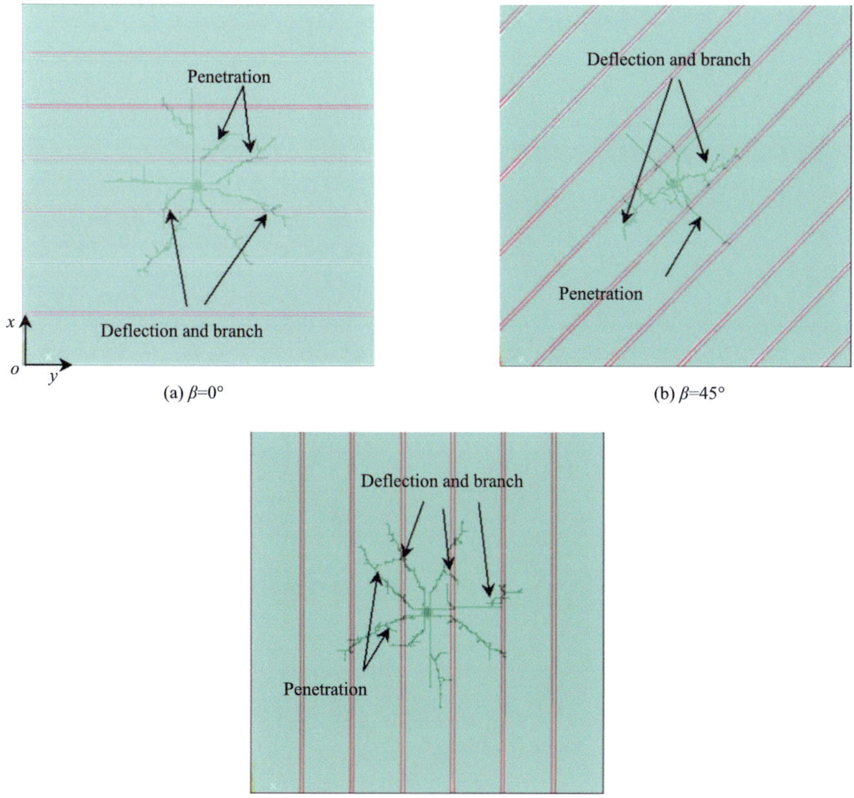

(a) β=0° (b) β=45° (c) β=90°

Fig. 3.8 Deflection behaviours and morphology of hydraulic fractures meeting beddings with variable geometrical configurations in Case II

hydraulic fracture propagation for the cases of $\beta = 0°$ and 90° is somewhat different.

(b) The fractal dimension for the cases of $\beta = 45°$ is smaller than that under $\beta = 0°$ and 90°. This is because the hydraulic fracture propagation is influenced by the bedding plane, resulting in deflection and propagation along the bedding plane, which affects the degree and complexity of hydraulic fracture network.

(c) The fractal dimensions in Case I with beddings under various bedding angles ($\beta = 0°$, 45°, and 90°) are the highest, followed by Case II, and the lowest in Case III, which is because the comprehensive enhanced geomechanical properties in bedding geomaterials of the bedding planes hinder the propagation of hydraulic fractures, leading to a decrease in the complexity of the hydraulic fracture network and the fractal dimension.

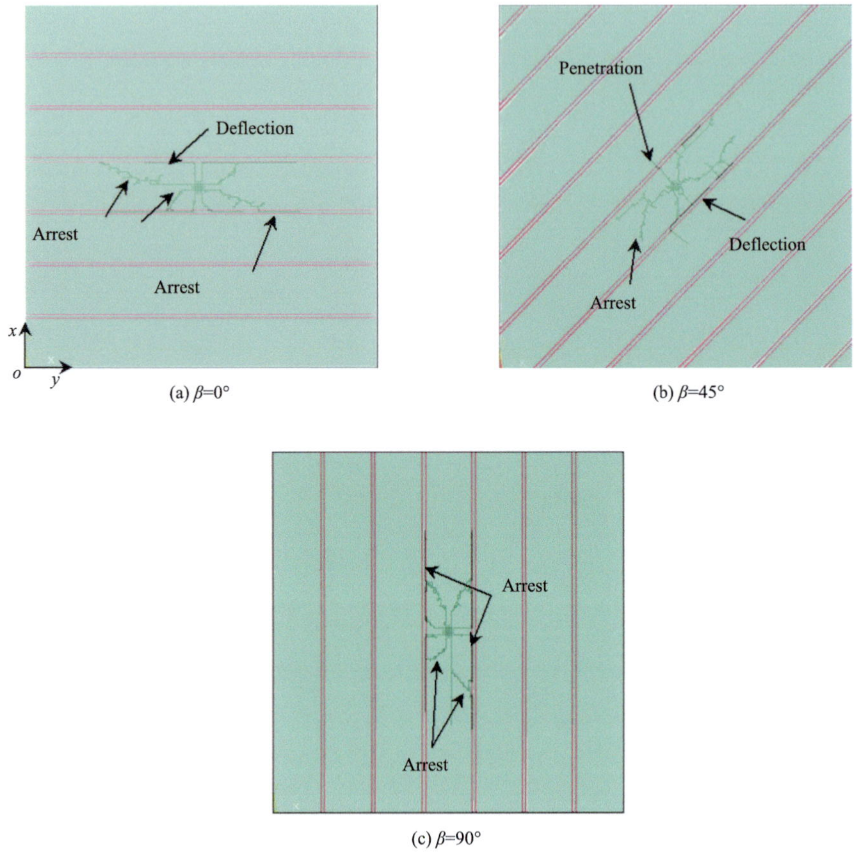

Fig. 3.9 Deflection behaviours and morphology of hydraulic fractures meeting beddings with variable geometrical configurations in Case III

Fig. 3.10 Fractal dimensions for morphology of hydraulic fractures meeting beddings with variable geometrical configurations

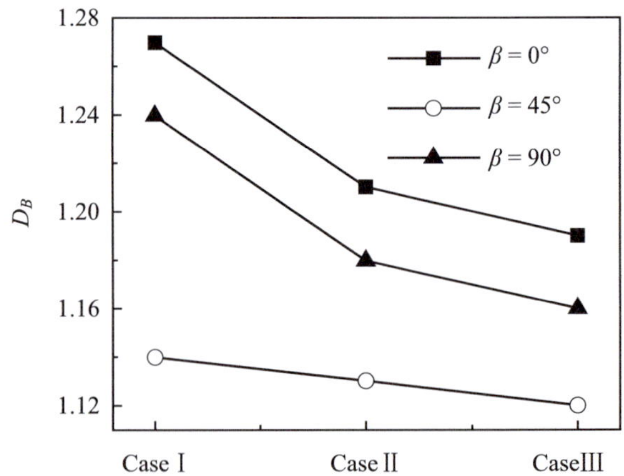

3.6.2 Beddings with Variable Geomechanical Properties

For better comparison, the granules model under bedding angles $\beta = 0°$ was assigned different geomechanical properties for computation and analysis. Figure 3.11 shows the results of deflection behaviours and morphology of hydraulic fractures meeting beddings with different Young's modulus; it can be seen that when bedding planes take small values, hydraulic fractures are easy to penetrate beddings; as the values continue to increase, hydraulic fractures tend to deflect and branch, forming a more complex network of hydraulic fractures.

Figures 3.12 and 3.13 show the results of deflection behaviours and morphology of hydraulic fractures meeting beddings with different tensile strength and cohesion, respectively, which reveals that the larger the tensile strength and cohesion are, the less the number of deflections of fractures is, the simpler the morphology is; the tensile strength and cohesion increase the strength of the bedding planes, which leads

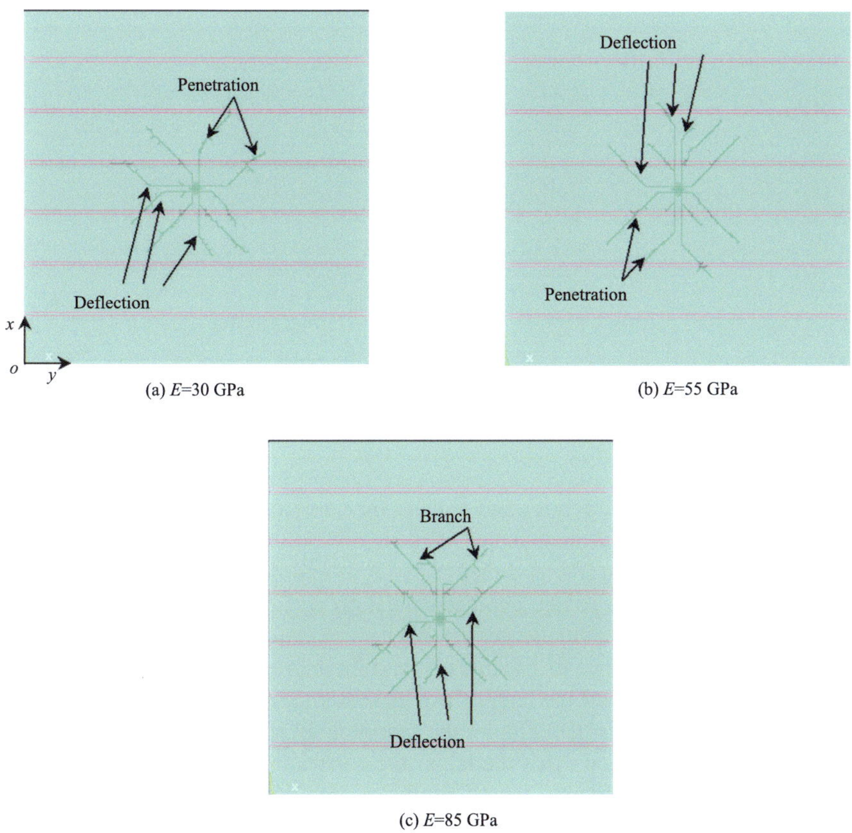

Fig. 3.11 Deflection behaviours and morphology of hydraulic fractures meeting beddings with different Young's modulus

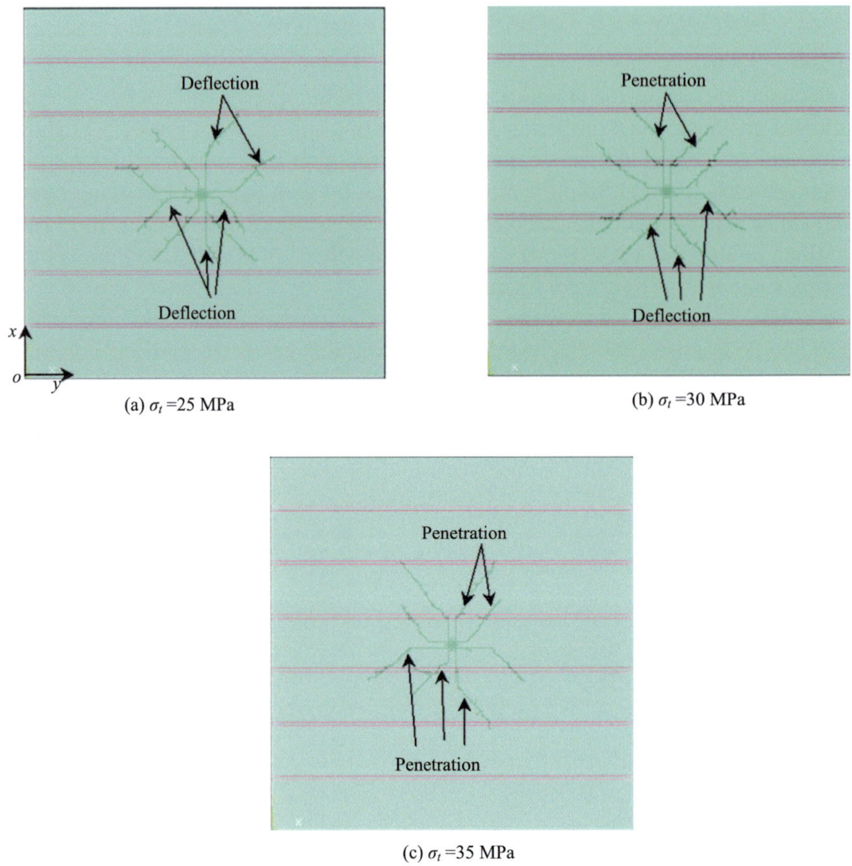

Fig. 3.12 Deflection behaviours and morphology of hydraulic fractures meeting beddings with different tensile strength

to many penetrations of fractures and reduces the complexity and fractal dimension of fracture network.

Figure 3.14 shows the fractal dimensions for morphology of hydraulic fractures meeting beddings with variable geomechanical properties in Cases IV, V, and VI. In Fig. 3.14a, it shows that the larger the Young's modulus is, the more the number of deflections and branches of fractures is, and the more complex the morphology is. Figure 3.14b shows that the larger the tensile strength is, the less the number of deflections of fractures is, the simpler the morphology is, so the fractal dimension is smaller; Fig. 3.14c shows that the larger the cohesion is, the less the number of deflections of fractures and more penetration arise, the simpler the fractal morphology is, so the fractal dimension of the fracture is smaller; the tensile strength and cohesion increase the strength of the bedding planes, which leads to many penetrations of fractures and reduces the complexity and fractal dimension of fracture network.

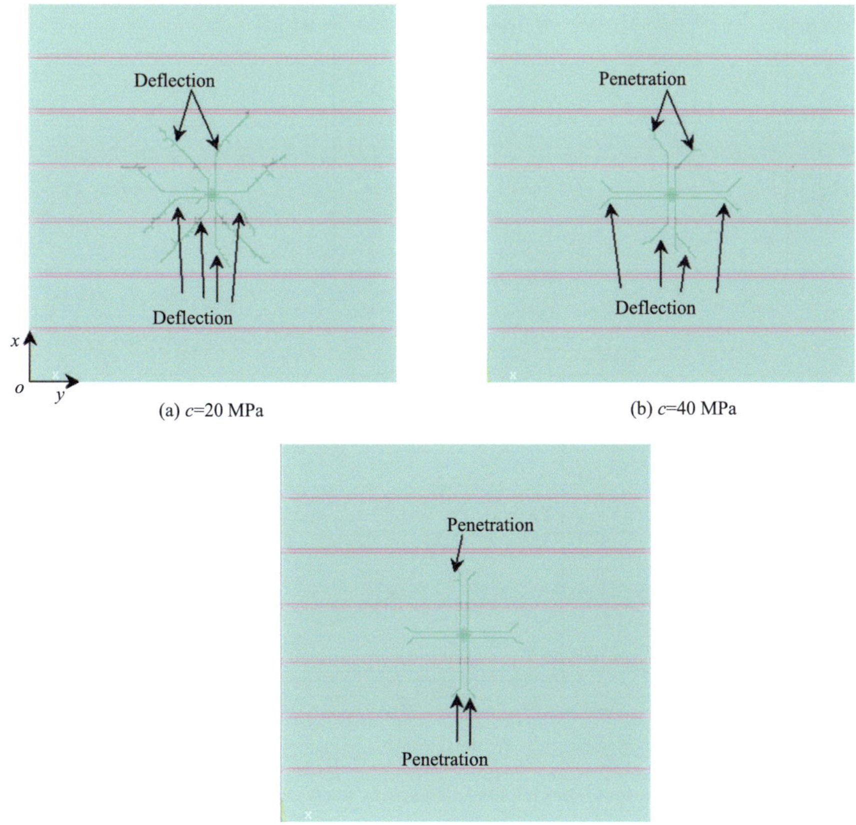

(a) c=20 MPa

(b) c=40 MPa

(c) c=60 MPa

Fig. 3.13 Deflection behaviours and morphology of hydraulic fractures meeting beddings with different cohesion

3.6.3 Granules with Variable Geometrical Configurations

Figures 3.15, 3.16 and 3.17 show the results of deflection behaviours and morphology of hydraulic fractures meeting granules with variable geometrical configurations, using the geomechanical properties of Case I II, and III. For the results in Configurations A, B, and C, the deflection, penetration, and arrest behaviours of hydraulic fractures arise. In Configuration A (Figs. 3.15a, 3.16a, and 3.17a), the deflections and braches of fractures arise induced by the small granule size; the small granule size increases the probability of fracture deflection and complexity of the fracture network. Once the granule size increases, the chance of fractures encountering granules decreases, making it easier for fractures to propagate straight and less prone to fracture branching, as shown in Figs. 3.15c, 3.16c, and 3.17c.

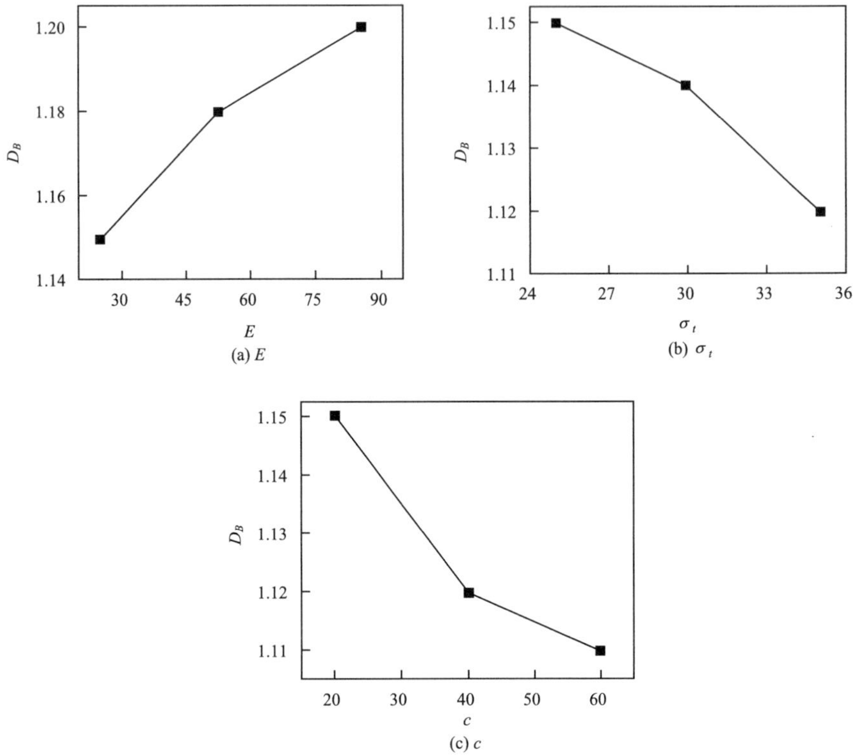

Fig. 3.14 Fractal dimensions for morphology of hydraulic fractures meeting beddings with variable geomechanical properties

Figure 3.18 shows the fractal dimensions for morphology of hydraulic fractures meeting granules with variable geometrical configurations, in which the following conclusions can be drawn:

(a) The fractal dimension of the fracture network under Configuration A is the largest, followed by Configuration B, and the smallest is under Configuration C. The smaller the granule size of different granule configurations under the same geomechanical properties, the larger the fractal dimension, indicating that a smaller granule size increases the probability of fracture deflection and complexity of the fracture network.
(b) The fractal dimensions in Case I under granules with variable geometrical configurations (Configurations A, B, and C) are the highest, followed by Case III, and the lowest in Case II, which is because the enhanced geomechanical properties (Young's modulus, tensile strength, and cohesion) in granule geomaterials (Cases II and III) hinder the propagation of hydraulic fractures, leading to a decrease in the complexity of the hydraulic fracture network and the fractal dimension comparing to the Case I; this is similar to the impact of heterogeneity

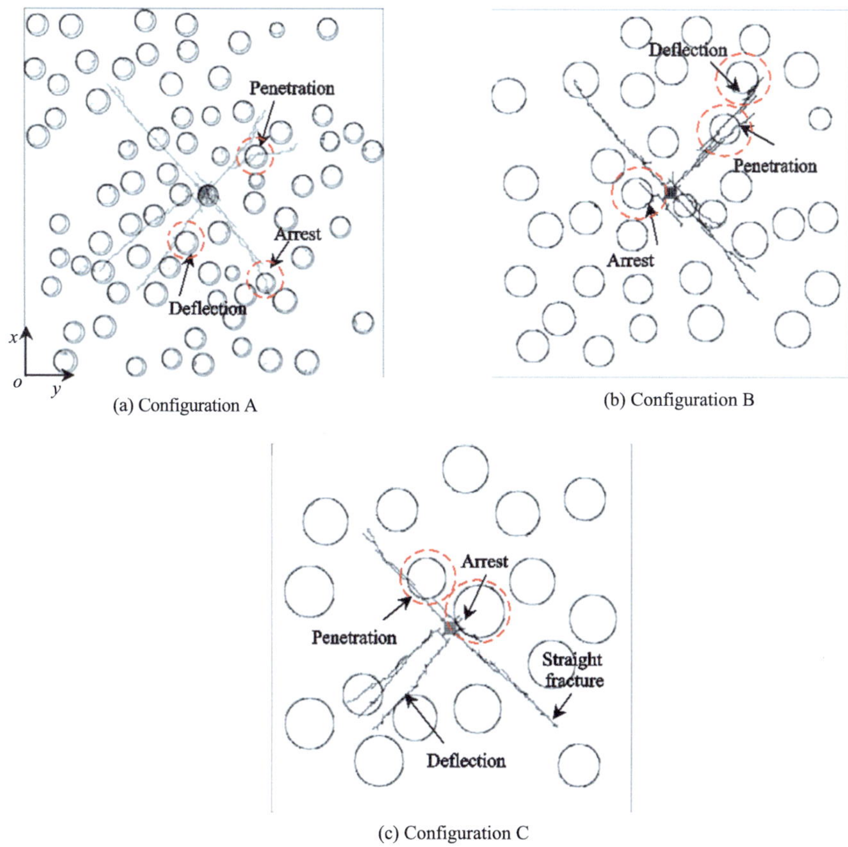

Fig. 3.15 Deflection behaviours and morphology of hydraulic fractures meeting granules with variable geometrical configurations in Case I

of bedding planes. From Case II to Case III, as the geomechanical properties of the granules increase for improving their stiffness and strength, it can be seen that the fractures are prone to be influenced by the granules, resulting in many deflections and branches and larger fractal dimension.

3.6.4 Granules with Variable Geomechanical Properties

For better comparison, the granules model under Configuration A was assigned different geomechanical properties for computation and analysis. Figures 3.19, 3.20, and 3.21 show the results of deflection behaviours and morphology of hydraulic fractures meeting granules with different Young's modulus, tensile strength, and cohesion, respectively. It can be seen that when granules have small values of geomechanical properties (Young's modulus, tensile strength, and cohesion), hydraulic fractures

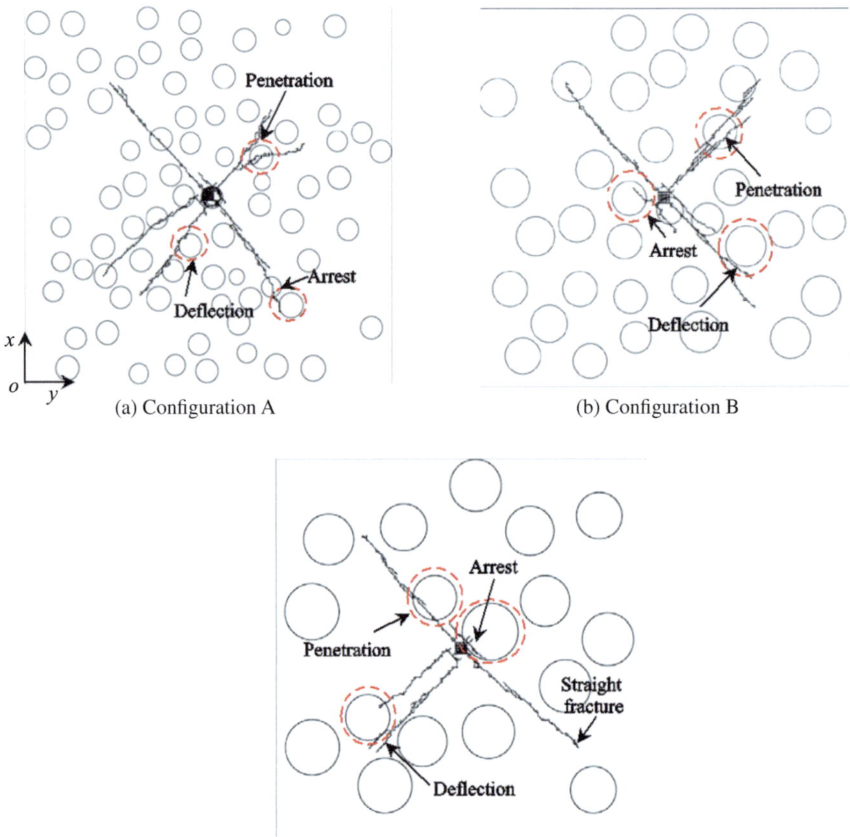

(a) Configuration A
(b) Configuration B
(c) Configuration C

Fig. 3.16 Deflection behaviours and morphology of hydraulic fractures meeting granules with variable geometrical configurations in Case II

are easy to penetrate granules, and the fractures almost keep straight; once the values increase, hydraulic fractures are prone to deflection around granules; as the values continue to increase, hydraulic fractures tend to deflect, branch, and arrest around granules, forming a more complex network of hydraulic fractures.

Figure 3.22 shows the fractal dimensions for morphology of hydraulic fractures meeting granules with variable geomechanical properties in Cases IV, V, and VI. As the geomechanical properties (Young's modulus, tensile strength, and cohesion) of the granules increase for improving their stiffness and strength, it can be seen that the fractures are prone to be influenced by the granules, resulting in many deflections and branches and larger fractal dimension, which are consistent with the multi-scale coupled processes modelling of fractures as porous, interfacial and granular systems (Hu & Rutqvist, 2021).

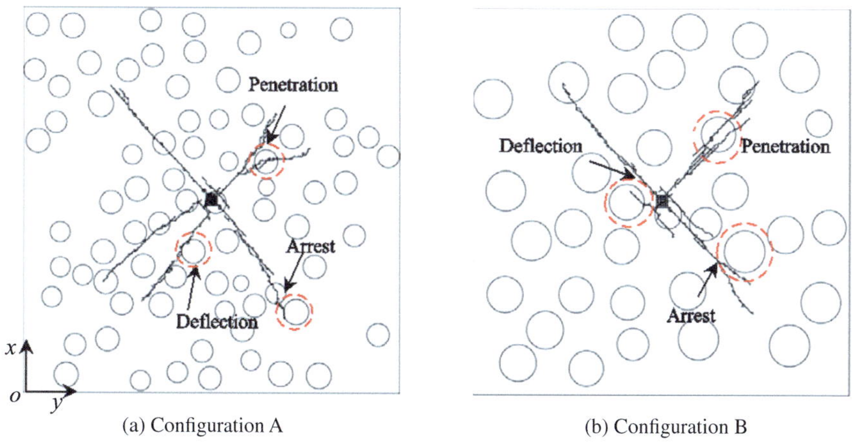

Fig. 3.17 Deflection behaviours and morphology of hydraulic fractures meeting granules with variable geometrical configurations in Case III

3.7 Conclusions

The main conclusions of this study are as follows:

(1) To investigate the deflection behavior and fractal morphology of hydraulic fractures meeting bedding and granules, the numerical models and cases with different geometrical configurations and geomechanical properties are proposed. Based on the combined finite element-discrete element-finite volume method and fractal characterization method, the fracture deflection and quantitative fractal morphology of hydraulic fractures considering the influences of beddings and granules are investigated and analyzed.

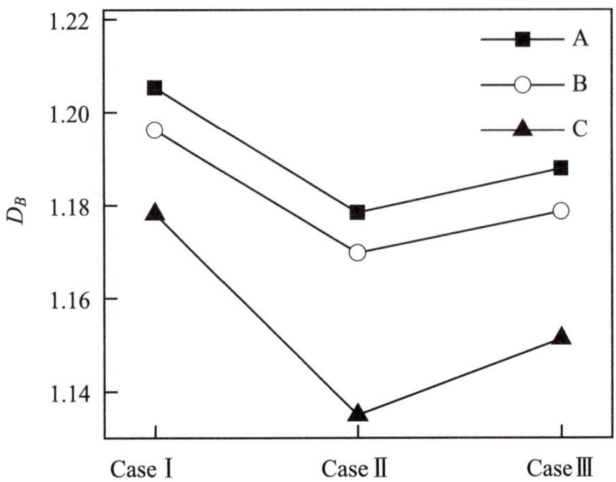

Fig. 3.18 Fractal dimensions for morphology of hydraulic fractures meeting granules with variable geometrical configurations

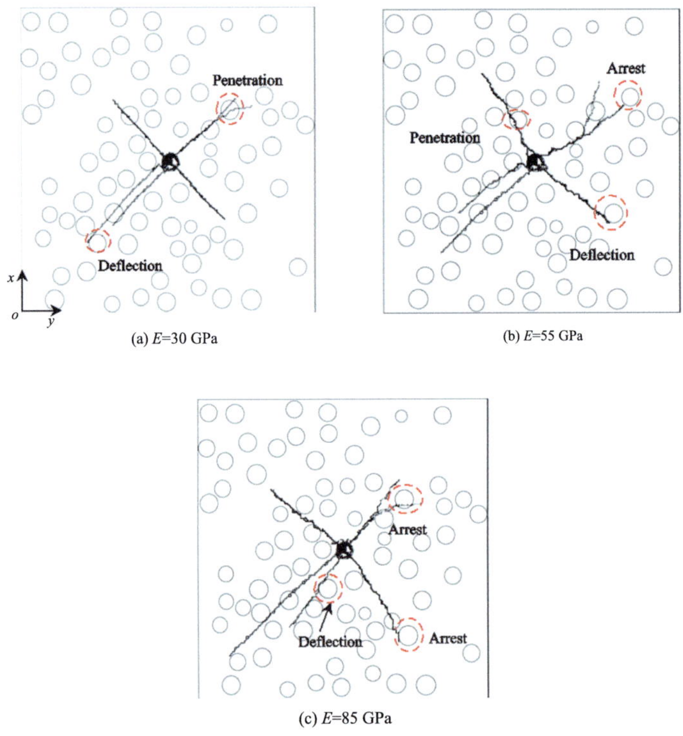

Fig. 3.19 Deflection behaviours and morphology of hydraulic fractures meeting granules with different Young's modulus

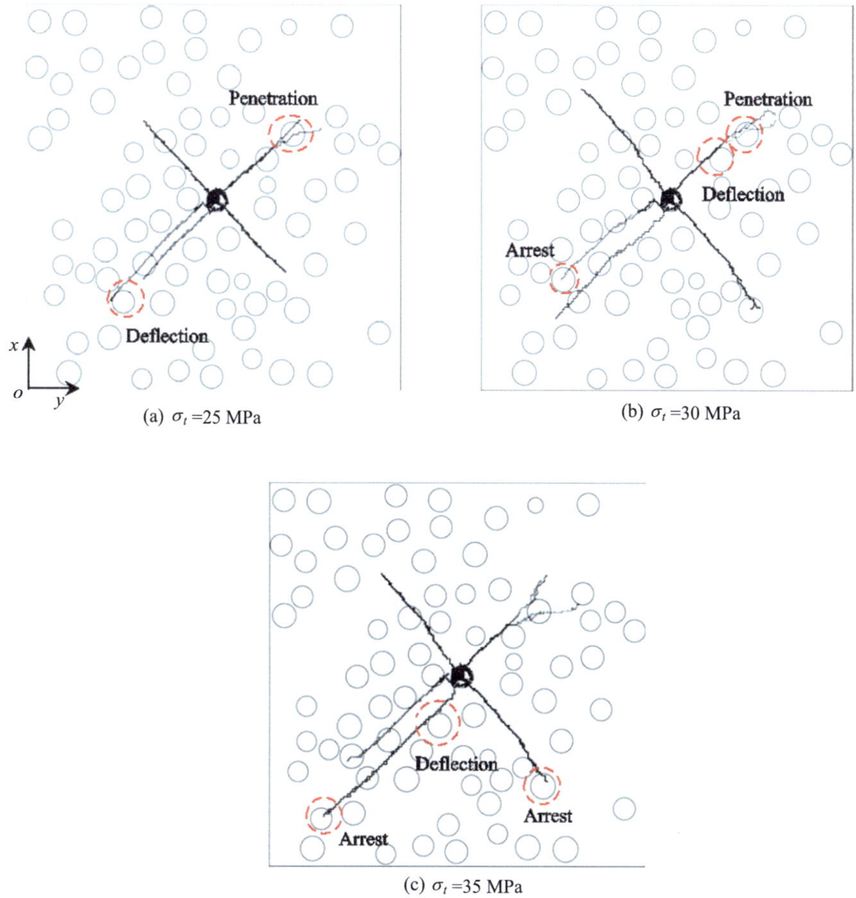

Fig. 3.20 Deflection behaviours and morphology of hydraulic fractures meeting granules with different tensile strength

(2) The fractal dimension for the cases of bedding dip angle $\beta = 45°$ is smaller than that under $\beta = 0°$ and 90°; this is because the hydraulic fracture propagation is influenced by the bedding plane, resulting in deflection and propagation along the bedding plane, which affects the degree and complexity of hydraulic fracture network. The comprehensive enhanced geomechanical properties (Young's modulus, tensile strength, and cohesion) in bedding geomaterials of the bedding planes hinder the propagation of hydraulic fractures, leading to a decrease in the complexity of the hydraulic fracture network and the fractal dimension; in details, the larger the Young's modulus is, the more the number of deflections and branches of fractures is, and the more complex the morphology is; the tensile strength and cohesion increase the strength of the bedding planes, which

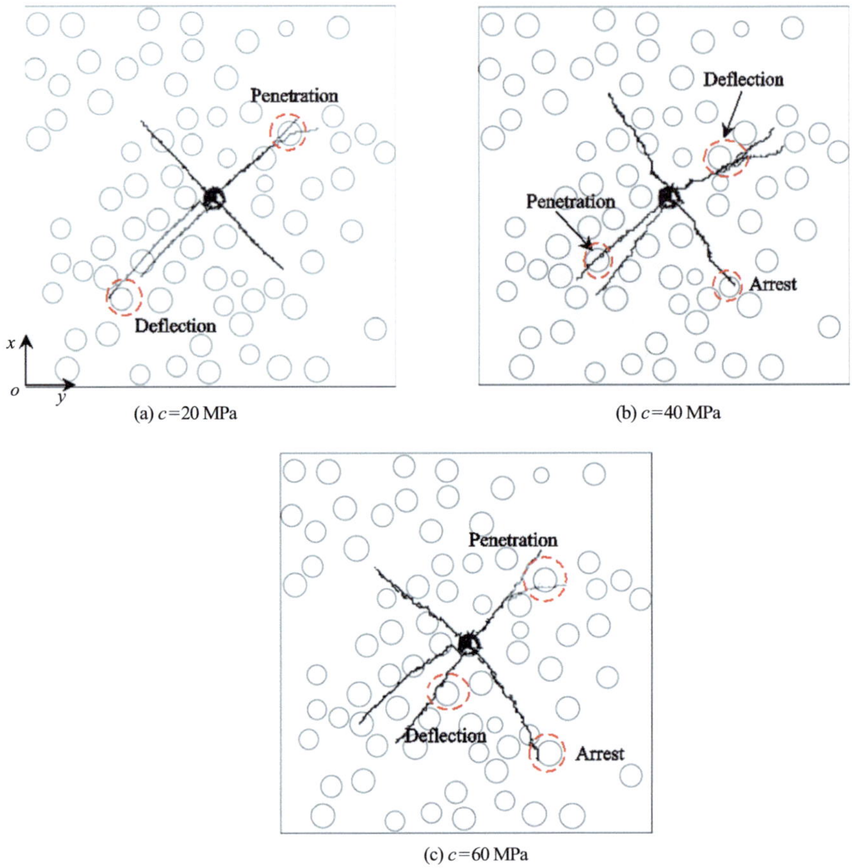

Fig. 3.21 Deflection behaviours and morphology of hydraulic fractures meeting granules with different cohesion

leads to many penetrations of fractures and reduces the complexity and fractal dimension of fracture network.

(3) The smaller the granule size of different granule configurations under the same geomechanical properties, the larger the fractal dimension, indicating that the small granule size increases the probability of fracture deflection and complexity of the fracture network. As the geomechanical properties (Young's modulus, tensile strength, and cohesion) of the granules increase for improving their stiffness and strength, it can be seen that the fractures are prone to be influenced by the granules, resulting in many deflections and branches and larger fractal dimension.

The investigation conclusions of fluid-driven fracture propagation with different geometrical configurations and geomechanical properties of beddings and granules can provide theoretical guidance for efficient and accurate evaluation of fracture

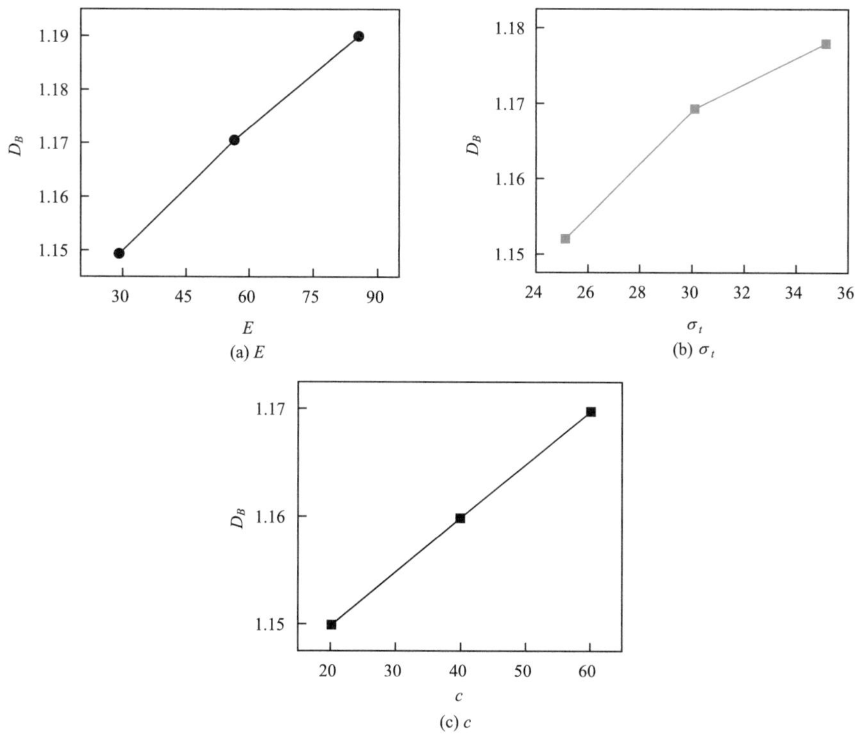

Fig. 3.22 Fractal dimensions for morphology of hydraulic fractures meeting granules with variable geomechanical properties

propagation behaviours and control and optimization of fracturing effects in practical engineering in heterogeneous reservoirs. This study is based on a two-dimensional model to evaluate the in-plane deflection and fractal morphology of hydraulic fractures, and the next stage of investigation will focus on the propagation behaviours of hydraulic fractures in three-dimensional heterogeneous reservoirs rock.

References

Bower, K. M., & Zyvoloski, G. (1997). A numerical model for thermo-hydro-mechanical coupling in fractured rock. *International Journal of Rock Mechanics and Mining Sciences, 34*(8), 1201–1211. https://doi.org/10.1016/S1365-1609(97)80071-8

Chen, W., Konietzky, H., Liu, C., & Tan, X. (2018). Hydraulic fracturing simulation for heterogeneous granite by discrete element method. *Computers and Geotechnics, 95*, 1–15. https://doi.org/10.1016/j.compgeo.2017.11.016

Chen, B., Barboza, B. R., Sun, Y., Bai, J., Thomas, H. R., Dutko, M., & Li, C. (2021). A review of hydraulic fracturing simulation. *Archives of Computational Methods in Engineering, 29*, 1–58. https://doi.org/10.1007/s11831-021-09653-z

Fu, P., Johnson, S. M., & Carrigan, C. R. (2013). An explicitly coupled hydro-geomechanical model for simulating hydraulic fracturing in arbitrary discrete fracture networks. *International Journal for Numerical and Analytical Methods in Geomechanics, 37*(14), 2278–2300. https://doi.org/10.1002/nag.2135

Gao, Q., & Ghassemi, A. (2020). Finite element simulations of 3D planar hydraulic fracture propagation using a coupled hydro-mechanical interface element. *International Journal for Numerical and Analytical Methods in Geomechanics, 44*(15), 11999–12024. https://doi.org/10.1002/nag.3116

Giovanardi, B., Serebrinsky, S., & Radovitzky, R. (2020). A fully-coupled computational framework for large-scale simulation of fluid-driven fracture propagation on parallel computers. *Computer Methods in Applied Mechanics and Engineering, 372*, Article 113365. https://doi.org/10.1016/j.cma.2020.113365

Han, Z., Zhou, J., & Zhang, L. (2018). Influence of grain size heterogeneity and in-situ stress on the hydraulic fracturing process by PFC2D modelling. *Energies, 11*(6), 1413. https://doi.org/10.3390/en11061413

Hu, M., & Rutqvist, J. (2021). Multi-scale coupled processes modeling of fractures as porous, interfacial and granular systems from rock images with the numerical manifold method. *Rock Mechanics and Rock Engineering, 55*, 3041–3059. https://doi.org/10.1007/s00603-021-02455-6

Li, M., Tang, S., Guo, T., & Qi, M. (2018). Numerical investigation of hydraulic fracture propagation in the glutenite reservoir. *Journal of Geophysics and Engineering, 15*(5), 2124–2138. https://doi.org/10.1088/1742-2140/aaba27

Li, S., Liu, L., Chai, P., Li, X., He, J., Zhang, Z., & Wei, L. (2019). Imaging hydraulic fractures of shale cores using combined positron emission tomography and computed tomography (PET-CT) imaging technique. *Journal of Petroleum Science and Engineering, 182*, Article 106283. https://doi.org/10.1016/j.petrol.2019.106283

Liang, X., Hou, P., Xue, Y., Gao, Y., Gao, F., Liu, J., & Dang, F. (2022). Role of fractal effect in predicting crack initiation angle and its application in hydraulic fracturing. *Rock Mechanics and Rock Engineering, 55*(9), 5491–5512. https://doi.org/10.1007/s00603-022-02940-6

Lisjak, A., Kaifosh, P., He, L., Tatone, B. S. A., Mahabadi, O. K., & Grasselli, G. (2017). A 2D, fully-coupled, hydro-mechanical, FDEM formulation for modelling fracturing processes in discontinuous, porous rock masses. *Computers and Geotechnics, 81*, 1–18. https://doi.org/10.1016/j.compgeo.2016.07.009

Liu, G., Cai, M., & Huang, M. (2018). Mechanical properties of brittle rock governed by micro-geometric heterogeneity. *Computers and Geotechnics, 104*, 358–372. https://doi.org/10.1016/j.compgeo.2017.11.013

McClure, M. W., Babazadeh, M., Shiozawa, S., & Huang, J. (2016). Fully coupled hydromechanical simulation of hydraulic fracturing in 3D discrete-fracture networks. *Spe Journal, 21*(4), 1302–1320. https://doi.org/10.2118/173354-PA

Meng, Y., Jing, H., Liu, X., Yin, Q., Zhang, L., & Liu, H. (2021). Experimental and numerical investigation on the effect of bedding plane properties on fracture behaviour of sandy mudstone. *Theoretical and Applied Fracture Mechanics, 114*, Article 102989. https://doi.org/10.1016/j.tafmec.2021.102989

Nguyen, V. P., Stroeven, M., & Sluys, L. J. (2012). Multiscale failure modeling of concrete: Micromechanical modeling, discontinuous homogenization and parallel computations. *Computer Methods in Applied Mechanics and Engineering, 201*, 139–156. https://doi.org/10.1016/j.cma.2011.09.014

Shuai, W., Ying, X., Yanbo, Z., Xulong, Y., Peng, L., & Xiangxin, L. (2023). Effects of sandstone mineral composition heterogeneity on crack initiation and propagation through a microscopic analysis technique. *International Journal of Rock Mechanics and Mining Sciences, 162*, Article 105307. https://doi.org/10.1016/j.ijrmms.2022.105307

Song, Y., Li, Z., Jiang, Z., Qun, L., Dongdong, L., & Zhiye, G. (2017). Progress and development trend of unconventional oil and gas geological research. *Petroleum Exploration and Development, 44*(4), 675–685. https://doi.org/10.1016/S1876-3804(17)30077-0

Song, J., Huo, Z., Fu, G., Hu, M., Sun, T., Liu, Z., & Liu, L. (2020). Petroleum migration and accumulation in the Liuchu area of Raoyang Sag, Bohai Bay Basin, China. *Journal of Petroleum Science and Engineering, 192*, Article 107276. https://doi.org/10.1016/j.petrol.2020.107276

Su, Y., Zhang, Q., Wang, W., & Sheng, G. (2015). Performance analysis of a composite dual-porosity model in multi-scale fractured shale reservoir. *Journal of Natural Gas Science and Engineering, 26*, 1107–1118. https://doi.org/10.1016/j.jngse.2015.07.046

Sun, Y., Edwards, M. G., Chen, B., & Li, C. (2021). A state-of-the-art review of crack branching. *Engineering Fracture Mechanics, 257*, Article 108036. https://doi.org/10.1016/j.engfracmech.2021.108036

Wang, C., & Wang, J. G. (2021). Effect of heterogeneity and injection borehole location on hydraulic fracture initiation and propagation in shale gas reservoirs. *Journal of Natural Gas Science and Engineering, 96*, Article 104311. https://doi.org/10.1016/j.jngse.2021.104311

Wang, Y., & Zhang, X. (2022). Dual bilinear cohesive zone model-based fluid-driven propagation of multiscale tensile and shear fractures in tight reservoir. *Engineering Computations, 39*(10), 3416–3441. https://doi.org/10.1108/EC-01-2022-0013

Wang, Y., Wang, J., & Li, L. (2022). Dynamic propagation behaviors of hydraulic fracture networks considering hydro-mechanical coupling effects in tight oil and gas reservoirs: a multi-thread parallel computation method. *Computers and Geotechnics, 152*, 105016.

Xie, H. (2020). *Fractals in rock mechanics*. CRC Press. https://doi.org/10.1201/9781003077626.

Zhang, G., Sun, S., Chao, K., Niu, R., Liu, B., Li, Y., & Wang, F. (2019). Investigation of the nucleation, propagation and coalescence of hydraulic fractures in glutenite reservoirs using a coupled fluid flow-DEM approach. *Powder Technology, 354*, 301–313. https://doi.org/10.1016/j.powtec.2019.05.073

Zhang, Q., Zhang, X. P., Ji, P. Q., Zhang, H., Tang, X., & Wu, Z. (2020). Study of interaction mechanisms between multiple parallel weak planes and hydraulic fracture using the bonded-particle model based on moment tensors. *Journal of Natural Gas Science and Engineering, 76*, Article 103176. https://doi.org/10.1016/j.jngse.2020.103176

Zhou, D., Zhang, G., Wang, Y., & Xing, Y. (2018). Experimental investigation on fracture propagation modes in supercritical carbon dioxide fracturing using acoustic emission monitoring. *International Journal of Rock Mechanics and Mining Sciences, 110*, 111–119. https://doi.org/10.1016/j.ijrmms.2018.07.010

Zou, Y., Zhang, S., Zhou, T., Zhou, X., & Guo, T. (2016). Experimental investigation into hydraulic fracture network propagation in gas shales using CT scanning technology. *Rock Mechanics and Rock Engineering, 49*, 33–45. https://doi.org/10.1007/s00603-015-0720-3

Chapter 4
Dynamic Propagation of Tensile and Shear Fractures Induced by Impact Load in Rock Based on Dual Bilinear Cohesive Zone Model

4.1 Introduction

Mounting global demand for coal, oil, and gas resources is elevating the strategic significance of deep unconventional reserves. During deep coal mining and rock reservoir excavation, complex fracture networks are induced under geomechanical constraints (Gutierrez & Youn, 2015; Ji et al., 2021; Shin & Santamarina, 2019). Reservoir complexities (including solid–fluid coupling, dynamic impact loads, and in-situ stress environments) govern multifaceted fracture initiation and propagation behaviors (Jeffrey et al., 2009; Peng et al., 2017; Wang et al., 2022a, 2023), encompassing tensile (Mode I), shear (Mode II), tear (Mode III), and mixed fracture modes. Evolving excavation conditions generate progressively complex fracture manifestations such as: fluid-driven reservoir fractures, impact-induced surrounding rock damage, and mining-triggered slope failures with fault slip (Beugelsdijk et al., 2000; Daneshy, 1974; Fan & Zhang, 2014; Lamont & Jessen, 1963; Warpinski & Teufel, 1987). Crucially, mechanical disturbances from blasting generate impact loads that propagate dynamic rock fractures, compromising reservoir stability (Hamid et al., 2018; Meglis et al., 2005). These transient high-intensity loads further develop complex fracture networks, inducing dynamic disasters (Li et al., 2021; Zhu & Tang, 2006). Consequently, preemptive determination of fracture location, typology, density, and morphology is essential for secure exploitation of deep unconventional resources.

Experimental and field approaches have been employed to characterize fracture propagation patterns in rock masses (Bohloli & De Pater, 2006; Bunger et al., 2015). Results reveal that fracture paths are governed by tensile-shear stress interactions, with tensile hydraulic fractures aligning parallel to maximum principal stress orientations while shear fractures dominate initial propagation stages. Acoustic emission frequency characteristics have been analyzed to identify propagation mechanisms (Deng et al., 2018). However, persistent scale limitations in experimental setups and

Y. Wang, *Deflection of Fracturing Fracture Network Disturbed by Discontinuity and Multiple Fractures in Rock*, https://doi.org/10.1007/978-981-95-1569-1_4

temporal constraints in monitoring impede in-situ differentiation of fracture modes and real-time observation of dynamic propagation processes.

Advancements in computational modeling have enabled diverse numerical methods (including finite element, discrete element, phase field (Miehe & Mauthe, 2016), energy minimization (Vahab et al., 2021), and peridynamic approaches (Hu et al., 2012)) to simulate multi-modal fracture propagation. Complementary fracture criteria encompass: maximum tensile stress (Palaniswamy & Knauss, 1978), Mohr–Coulomb (Wang et al., 2020), stress intensity factor-based (Irwin et al., 1958), energy-based (Griffith et al., 1921), and cohesive zone formulations (Dugdale, 1960). Practically, integrated finite element frameworks couple flow-stress-damage analyses, employing tensile and Mohr–Coulomb criteria for damage quantification (Yang et al., 2004), while linear elasticity principles dictate fracture initiation when tip stresses exceed material strength (Al-Rubaie & Ben Mahmud, 2020). Phase field models simulate brittle/ductile fractures in homogeneous/heterogeneous media (Borden et al., 2014), with parallel developments in peridynamic (Ren et al., 2016) and phase field methodologies (Miehe et al., 2010). Whereas conventional stress-based criteria (tensile/Mohr–Coulomb) dominate, they neglect energy dissipation during fracture evolution. Energy-based thresholds require strain energy release to exceed fracture resistance work for propagation (Griffith, 1920). Recent mixed-mode criteria incorporating energy evolution mechanisms (Wang & Zhang, 2022; Wang et al., 2022b) leverage dual bilinear cohesive zone models to investigate hydraulic fracture dynamics. Building on this foundation, the present study examines impact-induced tensile/shear fracture propagation in rocks using an advanced dual bilinear cohesive zone framework.

This chapter is structured as follows: In Sect. 4.2, the governing partial differential equations for rock fracture induced by impact load are presented. Section 4.3 describes the fracture criteria based on dual bilinear cohesive zone model, and Sect. 4.4 summarizes the numerical discretization of finite elements. Section 4.5 introduces the detection and separation of discrete elements. Section 4.6 presents the global algorithm and process of this study. In Sect. 4.7, the verification of tensile and shear fractures induced by impact load in rock is implemented; subsequently, the dynamic propagation of fractures in rock disc and rock stratum induced are analysed by numerical simulation at the laboratory- and engineering scale. The concluding remarks are provided in Sect. 4.8.

4.2 Governing Partial Differential Equations for Rock Fracture Induced by Impact Load

According to the theory of continuum mechanics, the governing partial differential equations for rock fracture induced by impact load can be expressed as follows:

$$\nabla \cdot \boldsymbol{\sigma} = \rho \ddot{\boldsymbol{u}} - \boldsymbol{f}, \quad x, y, z \in \Omega \tag{4.1}$$

where $\boldsymbol{u}$ is the displacement vector, $\boldsymbol{\sigma}$ represents the stress field vector, $\boldsymbol{f}$ is the external load vector (impact load), ρ is the density, $\ddot{\boldsymbol{u}}$ denote the acceleration vector, and Ω is the solution domain.

The strain–displacement relationship is given as

$$\varepsilon = \frac{1}{2}(\nabla \boldsymbol{u} + (\nabla \boldsymbol{u})^T) \tag{4.2}$$

where $\boldsymbol{\varepsilon}$ is the strain tensor.

The constitutive relation law is given as

$$\boldsymbol{\sigma} = \boldsymbol{D} : \boldsymbol{\varepsilon} \tag{4.3}$$

where $\boldsymbol{D}$ is the elastic matrix.

The boundary conditions are given as.

$$\boldsymbol{u} = \overline{\boldsymbol{u}}, \text{ on } \Gamma_u \tag{4.4a}$$

$$\boldsymbol{\sigma} \cdot \boldsymbol{n} = \overline{\boldsymbol{\sigma}}, \text{ on } \Gamma_\sigma \tag{4.4b}$$

where the displacement $\overline{\boldsymbol{u}}$ is prescribed on the displacement boundary Γ_u, and the confining stress $\overline{\boldsymbol{\sigma}}$ is prescribed on the external force boundary, Γ_σ.

The initial conditions are given as.

$$\boldsymbol{u}(t = 0) = \overline{\boldsymbol{u}}^0, \text{ on } \Gamma_u^0 \tag{4.5a}$$

$$\boldsymbol{\sigma}(t = 0) \cdot \boldsymbol{n} = \overline{\boldsymbol{\sigma}}^0, \text{ on } \Gamma_\sigma^0 \tag{4.5b}$$

where the displacement $\overline{\boldsymbol{u}}^0$ is prescribed on the initial displacement boundary Γ_u^0, and the confining stress $\overline{\boldsymbol{\sigma}}^0$ is prescribed on the initial external force boundary, Γ_σ^0.

4.3 Fracture Criteria Based on Dual Bilinear Cohesive Zone Model

The dual bilinear cohesive criterion includes the tensile cohesive criterion and shear cohesive criterion (Wang & Zhang, 2022; Wang et al., 2022b). Each fracture criterion includes the evolution process of the stress–strain relationship in rock mass: in the initial stage, the stress and strain show a linear increasing trend driven by the impact load; as the stress continues to increase, the stress reaches tensile or shear strength (in Fig. 4.1, $\sigma_{\max}$ is tensile strength and $\tau_{\max}$ is shear strength), and the cohesion of rock

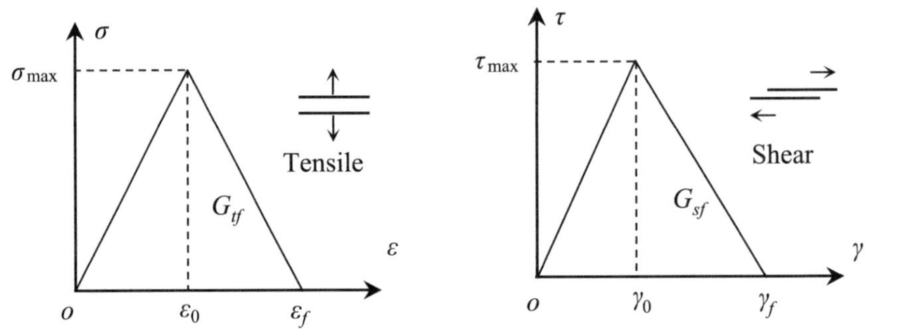

(a) Tensile fracture (Mode I fracture) (b) Shear fracture (Mode II fracture)

Fig. 4.1 Dual bilinear cohesive criteria for tensile and shear fracture of rock

material reaches its peak; subsequently, the rock mass begins to enter the stage of damage evolution (presented as plastic deformation), and the stress and strain show a linear decreasing trend; until the stress reduces to 0, the area surrounded by the stress–strain curve reaches the fracture energy of tensile or shear fracture of rock mass material (e.g., Equation (4.6a, 4.6b)); at this time, the rock mass reaches the maximum damage, that is, tensile or shear fracture occurs.

This failure process is that of energy dissipation of material cohesion, and the damage evolution and plastic deformation at the fracture tip shown in Fig. 4.1 can be effectively characterized. Mode I represents tensile fracture, and Mode II represents shear fracture, in Figs. 4.1a and b, respectively. In Fig. 4.1a, when $\varepsilon_0 < \varepsilon < \varepsilon_f$, the rock mass undergoes plastic deformation and enters the stage of damage evolution (softening stage); when $\varepsilon > \varepsilon_f$, the rock mass undergoes tensile fracture. The evolution of Fig. 4.1b was similar. The dual bilinear cohesive fracture criteria are expressed as:

$$\text{Tensile failure:} \quad \int_{tf} \sigma d\varepsilon = \int_{\varepsilon_0}^{\varepsilon_f} \sigma d\varepsilon = G_{tf} \tag{4.6a}$$

$$\text{Shear failure:} \quad \int_{sf} \tau d\gamma = \int_{\gamma_0}^{\gamma_f} \tau d\gamma = G_{sf} \tag{4.6b}$$

where σ is the tensile stress, τ is the shear stress, ε_0 is the tensile strain value when the tensile stress reaches the maximum value, ε_f is the tensile strain value when tensile fracture occurs, γ_0 is the shear strain value when the shear stress reaches the maximum value, γ_f is the shear strain value when shear fracture occurs, and G_{tf} and G_{sf} are the tensile fracture energy and shear fracture energy, respectively.

4.4 Numerical Discretization of Finite Elements

To solve the governing equations of solid deformation based on the FEM and using the variation formulation, the equilibrium Eq. (4.1) can be transformed into the following matrix form:

$$\boldsymbol{M}\ddot{\boldsymbol{D}} + \boldsymbol{K}\boldsymbol{D} = \boldsymbol{F}, \quad x, y, z \in \Omega \tag{4.7}$$

where $\boldsymbol{D}$ is the displacement vector; $\ddot{\boldsymbol{D}}$ denotes vector containing the nodal acceleration; $\boldsymbol{M}$ and $\boldsymbol{K}$ are the mass and stiffness matrices, respectively; $\boldsymbol{F}$ is the external vector.

4.5 Detection and Separation of Discrete Elements

Fracture propagation in this study is dynamically simulated through element boundary separation. The identification of separated element boundaries provides the quantitative foundation for determining fracture propagation direction and length. As impact loading pressure increases, inter-element contact forces satisfy the bilinear fracture criterion (Eq. (4.6a, 4.6b)), triggering separation at failed element interfaces. Figure 4.2 schematically illustrates tensile/shear nodal failure and fracture propagation under impact loading. The initial geometric domain (Fig. 4.2a) is discretized into finite elements (A, B, C, D) connected by nodes (Fig. 4.2b). When nodal stress/energy states meet dual bilinear cohesive fracture criteria (e.g., A-B interface node in Fig. 4.2c), initial fracture initiates via tensile/shear separation. Progressive loading induces further criterion-satisfying nodal failures (e.g., C-D node in Fig. 4.2d), extending fractures through continuous nodal separation. This nodal detection-separation mechanism thus implements impact-driven tensile/shear fracture propagation.

4.6 Global Algorithm and Procedure

Figure 4.3 illustrates the global algorithm and procedural workflow for simulating impact-induced tensile and shear fracture propagation using the dual bilinear cohesive zone model. The methodology initiates by transforming impact-loaded rock masses into mechanical models, subsequently forming corresponding finite element representations. Stress fields are computed through finite element analysis, enabling discrete element detection and separation via dual bilinear cohesive fracture criteria. As per these criteria, tensile and shear fractures initiate and propagate to generate updated fracture topologies and geometric configurations. Under progressive impact loading increments, fractures progressively propagate throughout the rock mass, ultimately forming interconnected tensile and shear fracture networks.

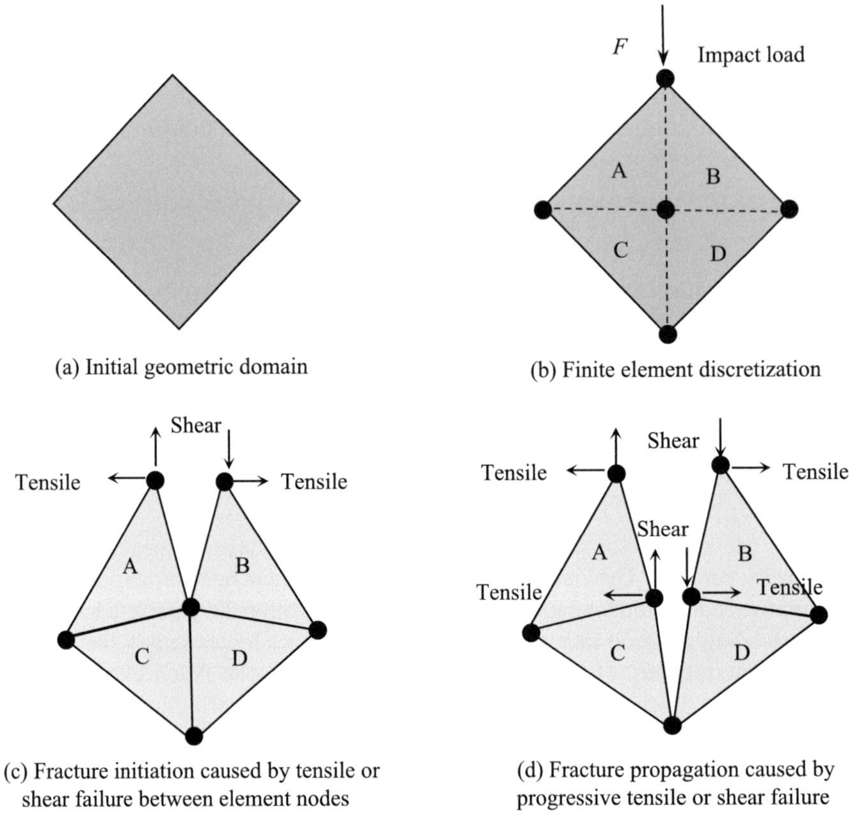

Fig. 4.2 Schematic of tensile or shear failure of nodes and fracture propagation induced by impact load

4.7 Results and Discussion

4.7.1 Verification of Tensile and Shear Fractures Induced by Impact Load in Rock Disc

To verify the tensile and shear fractures induced by impact load in rock based on the proposed method in this study, the failure modes of sandstone disc under impact load F in Brazilian disc splitting experiments (Li et al., 2023; Su et al., 2023) are provided as shown in Fig. 4.4, and the tensile and shear fractures arise in typically different regions. In the triangular regions where the upper and lower parts of the sandstone disc contact the loading force point, they are squeeze domains where many shear fractures arise; a fracture from the upper to lower parts will appear in the central domain of the sandstone disc, which is a typical tensile fracture caused by the increase of horizontal tensile stress in the central domain. The coexistence of tensile and shear fractures was tested in this experiment, and the fracture morphology and

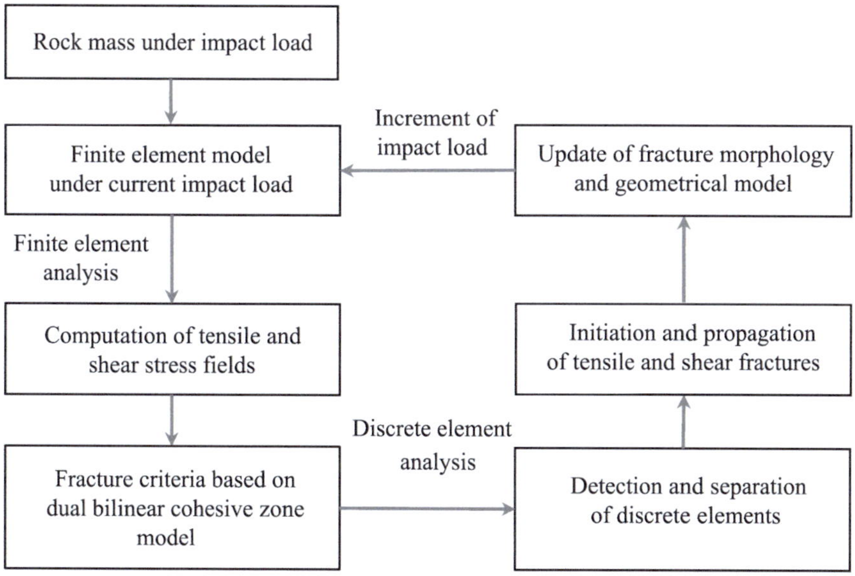

Fig. 4.3 Global program for dynamic propagation of impact tensile and shear fractures based on dual bilinear cohesive zone model

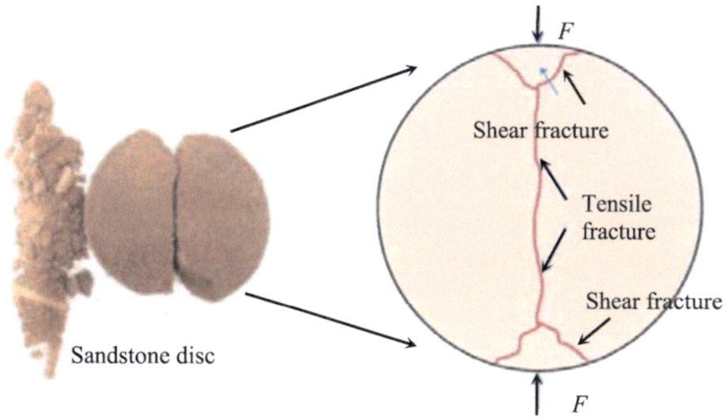

Fig. 4.4 Failure modes of sandstone disc under impact load in experiments

distribution region were obtained. These experimental results can be well used here to verify the effectiveness and reliability of the proposed method in this study.

To numerically analyze the process of fracture propagation under impact load, the finite element model of laboratory-scale rock disc under impact load F is established as shown in Fig. 4.5, and the physical parameters of rock disc under impact load is shown in Table 4.1. The diameter of the model is 100 mm, which is divided into 3000 triangular elements with a side length of $l = 2$ mm. The boundary conditions and load states are the same as the experiments, and the velocity of impact load is $v = 2$ m/s and total loading step is $n = 10{,}000$.

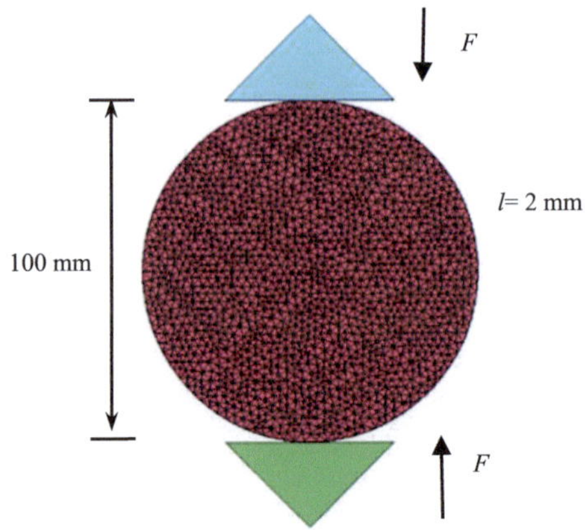

Fig. 4.5 Finite element model of laboratory-scale rock disc under impact load

Table 4.1 Physical parameters of rock disc under impact load

E/GPa	ν	ρ/(kg/m^3)	$\sigma_{\max}$/MPa	$\tau_{\max}$/MPa	G_{tf}/(N·m/m^2)	G_{sf}/(N·m/m^2)
69	0.22	2670	7.8	20	300	300

The computed results of rock disc under impact load are shown in Fig. 4.6, and the following conclusions can be drawn: (a) The tensile fractures initiate from the center of the model and propagate vertically along the direction of impact load; (b) The shear fractures are distributed in the triangular compression region near the impact load, and the short and concentrated fracture morphology is consistent with the experimental results (Li et al., 2023; Su et al., 2023); (c) Both tensile and shear fractures occurred in numerical model, and the type, distribution, and propagation process of fractures can be derived. Therefore, the effectiveness and reliability of the proposed method are well verified.

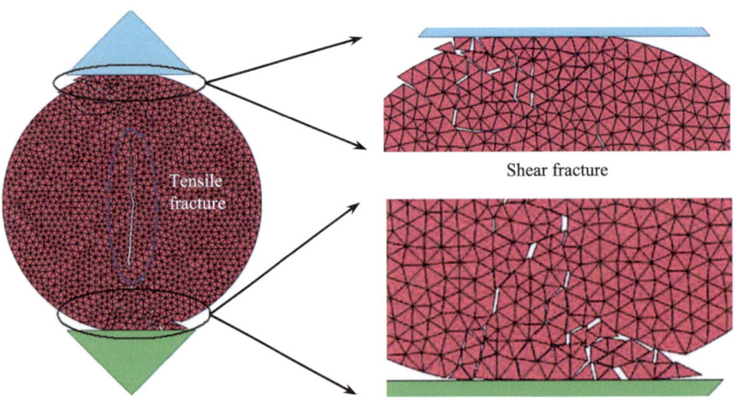

Fig. 4.6 Final fracture morphology of rock disc under impact load velocity $v = 2$ m/s

4.7.2 *Dynamic Propagation of Fractures in Rock Disc*

4.7.2.1 Laboratory-Scale Rock Disc Under Different Impact Load Steps

The computed results of laboratory-scale rock disc are shown in Fig. 4.7, and the following conclusions for the fracture propagation of rock disc under different impact load steps by the proposed method can be drawn: (a) As shown in Fig. 4.7a, in the initial stage of impact loading ($n = 2000$), the rock disc undergoes plastic deformation due to stress concentration around the loading region, and then shear fractures begin to initiate accompanied by slip between elements; (b) As shown in Fig. 4.7b, as the impact loading increases ($n = 4000$), the shear fractures gradually propagate, and a tensile fracture imitates in the central region; (c) As shown in Fig. 4.7c, as the impact loading increases ($n = 6000$), the shear fractures propagate and connect

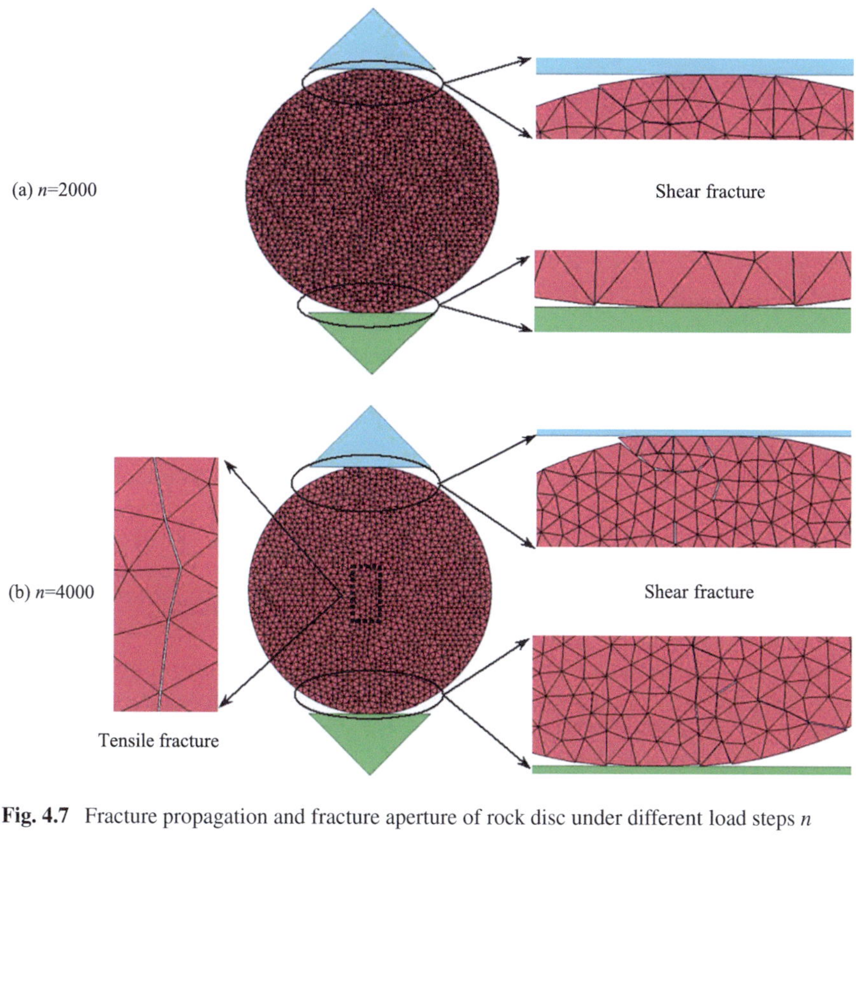

Fig. 4.7 Fracture propagation and fracture aperture of rock disc under different load steps n

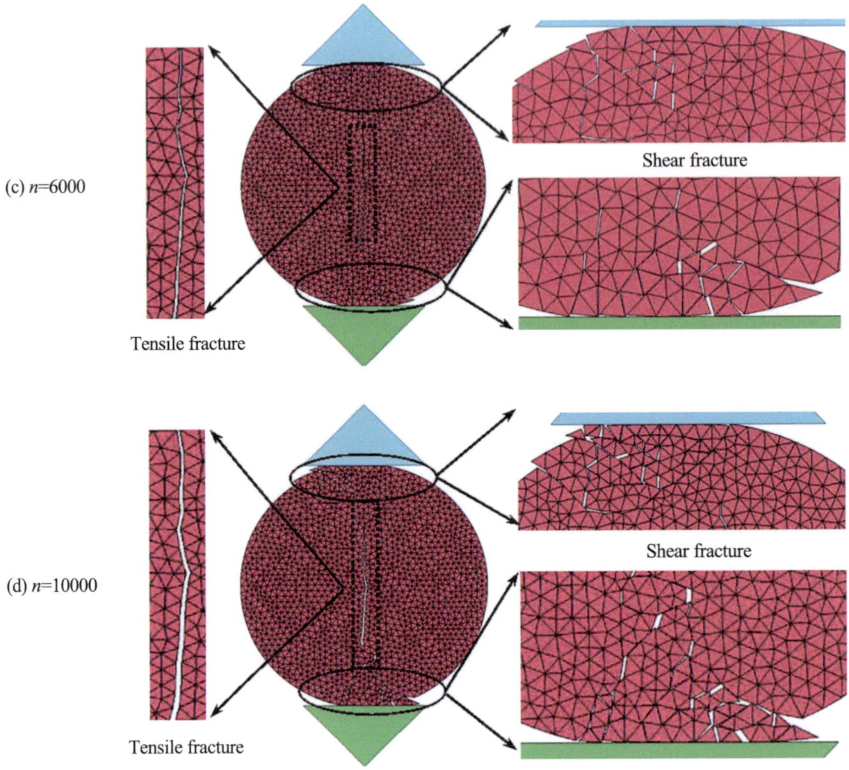

Fig. 4.7 (continued)

to cluster, and the tensile fracture vertically propagate and its width increases; (d) As shown in Fig. 4.7d, driven by the sustained impact load ($n = 10{,}000$), the shear fracture cluster continues to propagate, and the tensile fracture propagate to connect the shear fracture cluster; finally, the fractures penetrate the rock stratum, resulting in the entire failure.

4.7.2.2 Influence of Mesh Sensitivity on Fracture Morphology

To investigate the influence of mesh sensitivity on fracture morphology, the finite element model of laboratory-scale rock disc with different mesh density were established, as shown in Fig. 4.8. The mesh lengths were adopted as $l = 20$ mm (Fig. 4.8a) and $l = 5$ mm (Fig. 4.8b), respectively.

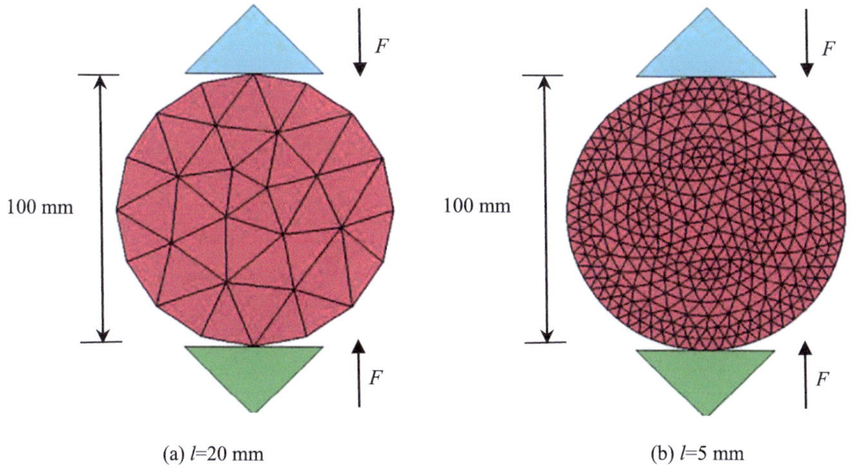

Fig. 4.8 Finite element model of laboratory-scale rock disc with different mesh density

The final tensile and shear fracture morphologies of laboratory-scale rock disc with different mesh density are shown in Figs. 4.9 and 4.10, respectively. As shown in Figs. 4.9a and 4.10a, when the element is relatively large ($l = 20$ mm), the tensile fracture may propagate along the edge of the element, resulting in its lack of fracture symmetry in the model; the shear fracture appears in the central regions, which deviated significantly from the experimental and numerical results mentioned above. As shown in Figs. 4.9b and 4.10b, when the element length is small ($l = 5$ mm), the tensile fractures arise in central region and shear fractures in upper and lower regions, respectively, which phenomenon are reliable due to the small element length. Therefore, in the above study, a smaller element length of $l = 2$ mm was used to guarantee the high-precision results.

4.7.2.3 Engineering-Scale Rock Disc Under Different Different Impact Load Velocities

To test the applicability of the proposed method in fracture propagation at different scales, the finite element model of engineering-scale rock disc under impact load F is established as shown in Fig. 4.11, and the physical parameters of rock disc under impact load is shown in Table 4.1 as above laboratory-scale case. The diameter of the model is 10 m, which is divided into 726 triangular elements with a side length of $l = 0.5$ m. The boundary conditions and load states are the same as the above laboratory-scale case, and the total loading step is $n = 25{,}000$.

The computed results of final fracture morphology of engineering-scale rock disc under different impact load velocities ($v = 20$ m/s, $v = 50$ m/s, $v = 100$ m/s) are

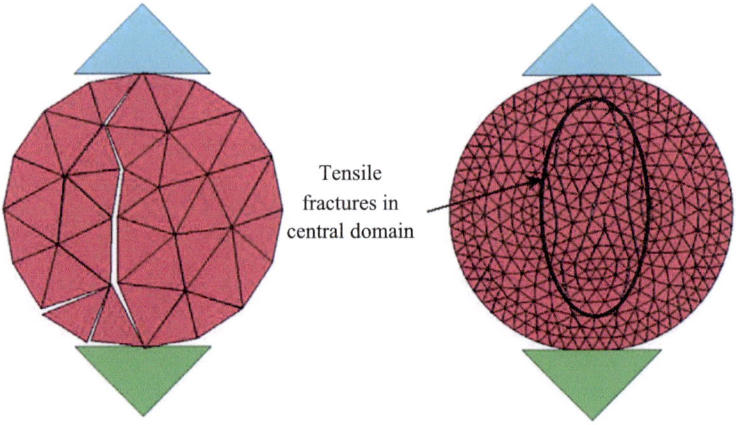

Fig. 4.9 Final tensile fracture morphology of laboratory-scale rock disc with different mesh density

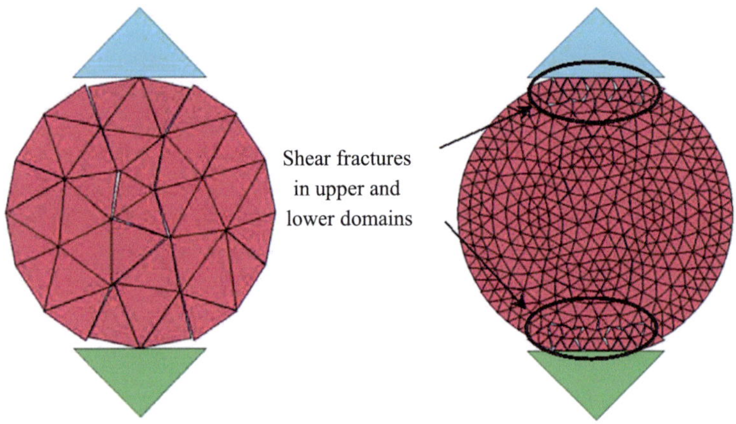

Fig. 4.10 Final shear fracture morphology of laboratory-scale rock disc with different mesh density

shown in Fig. 4.12, and the tensile and shear fractures appear. When the impact load velocity is small ($v = 20$ m/s), the fracture width is small, and the model maintains stability and integrity; once the velocity is high ($v = 100$ m/s), the fracture width is large, and there occurs entire failure. Therefore, the impact load velocity is an important factor inducing fracture; through the test, it can be seen that the proposed method has the application of the proposed method in fracture propagation at the engineering scale.

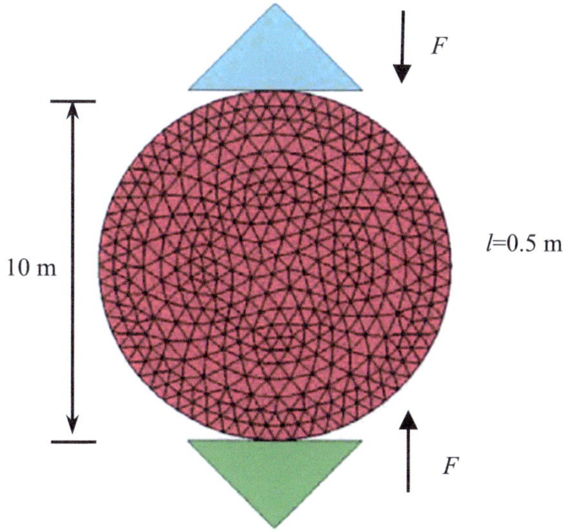

Fig. 4.11 Finite element model of engineering-scale rock disc under impact load

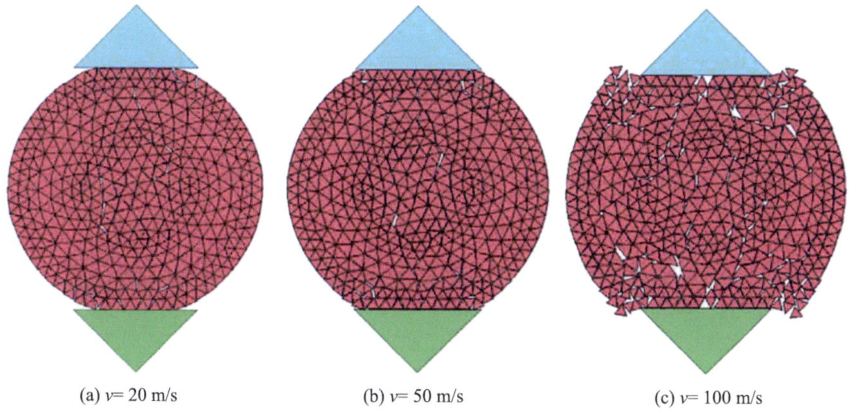

Fig. 4.12 Final fracture morphology of engineering-scale rock disc under different impact load velocities

4.7.3 Dynamic Propagation of Fractures in Rock Stratum

4.7.3.1 Laboratory-Scale Rock Stratum Under Different Impact Load Steps

Furthermore, the proposed method is extended to investigate the initiation and propagation of complex fractures of deep rock stratum under impact load. A finite element model of rock stratum under impact load F is shown in Fig. 4.13, and the physical parameters of rock stratum under impact load is shown in Table 4.2. The length and height of the model are 400 mm and 100 mm, respectively, which is divided into 1536 triangular elements with a side length of $l = 12.5$ mm. The bottom and left and right sides of the model are completely fixed, and the velocity of impact load is $v =$ 3 m/s and total loading step is $n = 12{,}000$.

The computed results of rock stratum under impact load are shown in Fig. 4.14, and the following conclusions for the fracture propagation of rock stratum under different impact load steps by the proposed method can be drawn: (a) As shown in Fig. 4.14a, in the initial stage of impact loading ($n = 3000$), the rock stratum undergoes plastic deformation due to stress concentration around the loading region, and then shear fractures begin to initiate; (b) As shown in Fig. 4.14b, as the impact loading increases ($n = 7500$), the shear fractures gradually propagate, showing a symmetrical distribution; the rarely transverse and radial fractures arise in the far-field region of the impact loading; (c) As shown in Fig. 4.14c, driven by the sustained impact load ($n = 12{,}000$), the fractures continue to propagate, forming several long fractures; finally, the fractures penetrate the rock stratum, resulting in the entire failure. From these results, the rock stratum exhibits local shear fracture clusters under impact load, which does not have a significant influence on the entire failure; as the impact load increases, several long fractures will initiate and propagate until the entire failure of rock stratum. Therefore, in the initial stage of impact loading, if there are reasonable monitoring results, the occurrence domain of impact load can be identified and early warning of entire failure can be made.

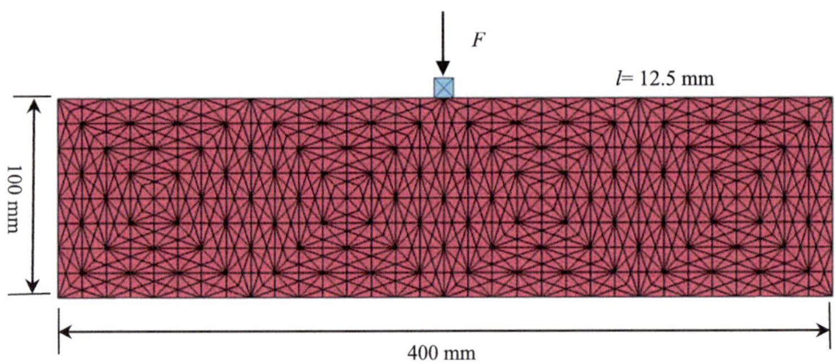

Fig. 4.13 Finite element model of rock stratum under impact load

Table 4.2 Physical parameters of rock stratum under impact load

E/GPa	v	ρ/(kg/m^3)	σ_{max}/MPa	τ_{max}/MPa	G_{tf} /(N·m/m^2)	G_{sf} /(N·m/m^2)
60	0.2	2500	6.9	16.7	250	250

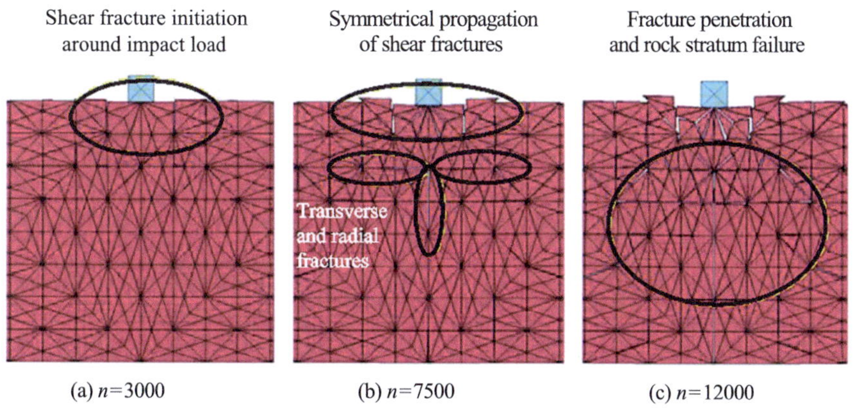

Fig. 4.14 Fracture propagation of rock stratum under different impact load steps by the proposed method

4.7.3.2 Engineering-Scale Rock Stratum Under Different Impact Load Velocities

To test the applicability of the proposed method in fracture propagation at different scales, the finite element model of engineering-scale rock stratum under impact load F is established as shown in Fig. 4.15, and the physical parameters of rock stratum under impact load is shown in Table 4.2 as above laboratory-scale case. The length and height of the model are 40 m and 10 m, respectively, which is divided into 1536 triangular elements with a side length of $l = 1.25$ m. To simulate the dynamic propagation of fractures induced by three-point bending of rock beams, the bottom left and right corners are fixed and the impact loading point is at the center of the upper side of the model, and the total loading step is $n = 12{,}000$.

The computed results of final fracture morphology of engineering-scale rock stratum under different impact load velocities ($v = 3$ m/s, $v = 5$ m/s, $v = 20$ m/s) are shown in Fig. 4.16, and the shear fractures are mainly concentrated in the

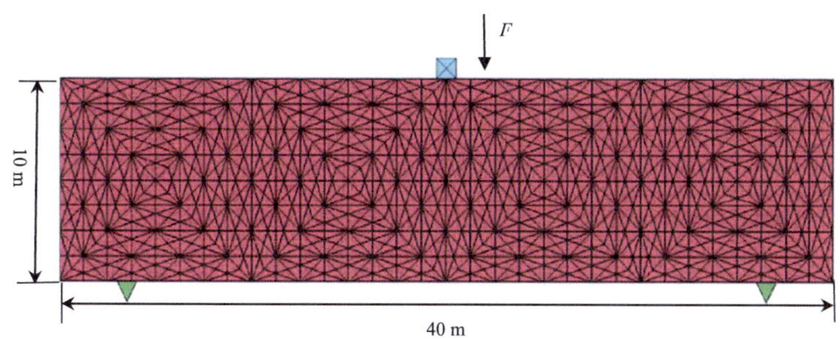

Fig. 4.15 Finite element model of engineering-scale rock stratum under impact load

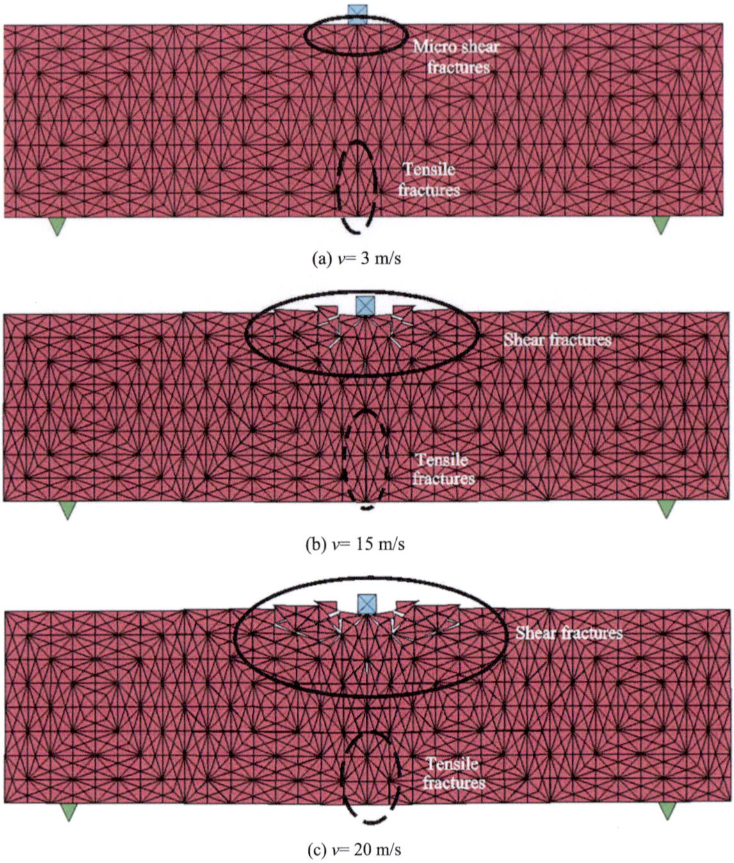

Fig. 4.16 Final fracture morphology of engineering-scale rock stratum under different impact load velocities

compression zone around the impact load, and the tensile fractures occur at the bottom of the model. When the impact load velocity is small ($v = 3$ m/s), the micro fracture width is small, and the model maintains stability and integrity; once the velocity is high ($v = 20$ m/s), the fracture width is large, and there occurs entire failure.

4.7.3.3 Engineering-Scale Rock Stratum Under Impact Load at Different Positions

To test the effects of impact loading point on fracture type and propagation morphology, the finite element model of engineering-scale rock stratum under impact load at different positions ($a = 9.5$ m, $a = 1.0$ m) is established as shown in Fig. 4.17. The velocity of impact load is $v = 3$ m/s, and the total loading step is $n = 12{,}000$.

The final fracture morphology of engineering-scale rock stratum under impact load at different positions is shown in Fig. 4.18. As shown in Fig. 4.18a, when the impact load deviates from the symmetrical position of rock stratum structure, the shear fracture propagates in an asymmetric manner. As shown in Fig. 4.18b, when the impact load is applied near the left support constraint boundary, concentrated shear fractures appear around the loading region, as well as induced shear fracture band, which may induce local instability.

4.8 Conclusions

This investigation yields the following principal findings:

(1) This study employs an advanced dual bilinear cohesive zone model coupled with a hybrid finite element-discrete element methodology to investigate impact-induced dynamic propagation of tensile and shear fractures in rock formations.

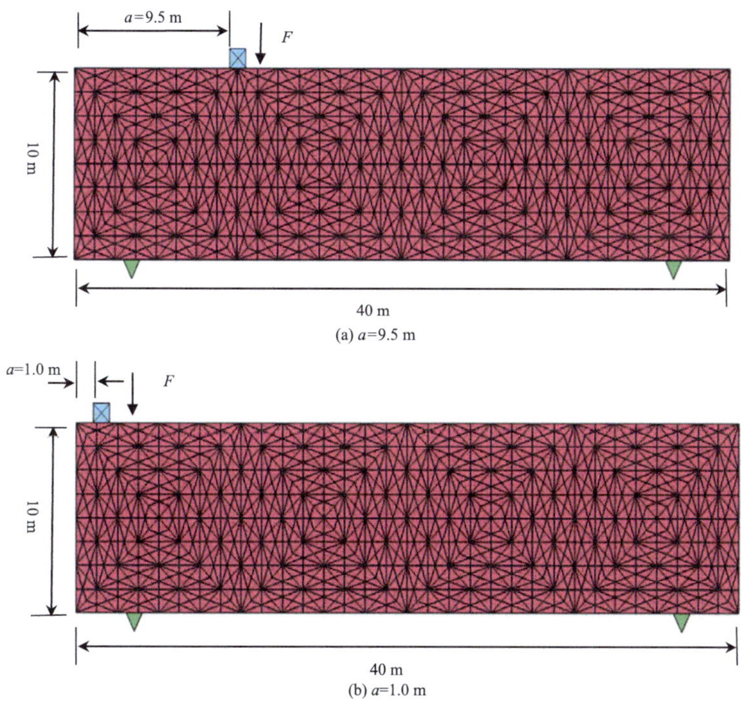

Fig. 4.17 Finite element model of engineering-scale rock stratum under impact load at different positions

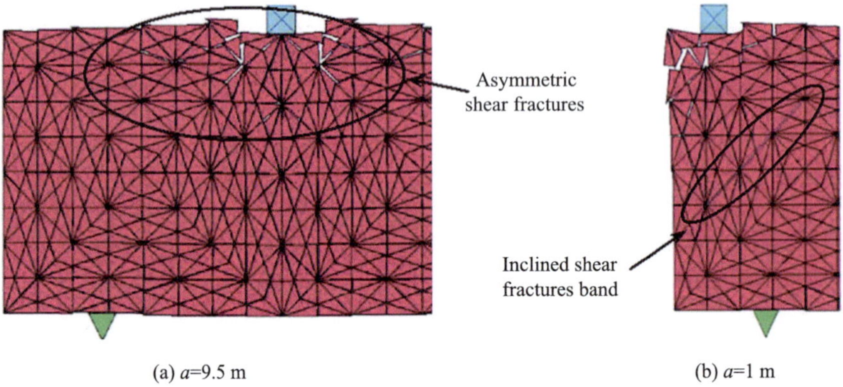

Fig. 4.18 Final fracture morphology of engineering-scale rock stratum (local region) under impact load at different positions

The implemented mixed-mode fracture criteria, incorporating energy evolution mechanisms, enable simultaneous simulation of both fracture types. Validation against experimental data from impact-loaded rock discs confirms the method's computational efficacy and reliability.

(2) Numerical simulations capture dynamic fracture propagation in laboratory-scale rock discs under incremental impact loading, delineating tensile and shear crack progression. Mesh sensitivity analysis demonstrates that optimized refinement ensures high-fidelity capture of fracture morphology. For engineering-scale discs, elevated impact velocities correlate with increased fracture widths and complete structural failure.

(3) The computational framework extends to impact-induced complex fracture initiation and propagation in deep rock strata across laboratory and engineering scales. Initially, localized shear fracture clusters develop without compromising structural integrity; subsequently, progressive loading triggers extensive fracture networks culminating in complete stratum failure. Parametric studies further demonstrate that elevated impact velocities accelerate failure with widened fractures, while near-boundary loading induces concentrated shear fractures and fracture bands causing local instability.

This research examines dynamic tensile and shear fracture propagation under impact loading in rock formations. The framework extends to investigating mixed-mode fractures and in-situ stress disturbances during dynamic strata mining in deep energy extraction, with future development of a dual bilinear cohesive zone model and simulation scheme for practical mining engineering applications.

References

Al-Rubaie, A., & Ben Mahmud, H. K. (2020). A numerical investigation on the performance of hydraulic fracturing in naturally fractured gas reservoirs based on stimulated rock volume. *Journal of Petroleum Exploration and Production Technology, 10*(8), 3333–3345.

Beugelsdijk, L. J. L., De Pater, C. J. and Sato, K. (2000), "Experimental hydraulic fracture propagation in a multi-fractured medium", *SPE Asia Pacific conference on integrated modelling for asset management*, SPE-59419-MS.

Bohloli, B., & De Pater, C. J. (2006). Experimental study on hydraulic fracturing of soft rocks: Influence of fluid rheology and confining stress. *Journal of Petroleum Science and Engineering, 53*(1–2), 1–12.

Borden, M. J., Hughes, T. J., Landis, C. M., & Verhoosel, C. V. (2014). A higher-order phase-field model for brittle fracture: Formulation and analysis within the isogeometric analysis framework. *Computer Methods in Applied Mechanics and Engineering, 273*, 100–118.

Bunger, A. P., Kear, J., Dyskin, A. V., & Pasternak, E. (2015). Sustained acoustic emissions following tensile crack propagation in a crystalline rock. *International Journal of Fracture, 193*(1), 87–98.

Daneshy, A. A. (1974). Hydraulic fracture propagation in the presence of planes of weakness. In *SPE Europec featured at EAGE Conference and Exhibition*, SPE-4852-MS.

Deng, J. H., Li, L. R., Chen, F., Liu, J. F., & Yu, J. (2018). Twin-peak frequencies of acoustic emission due to the fracture of marble and their possible mechanism. *Advanced Engineering Sciences, 50*(5), 1–6.

Dugdale, D. S. (1960). Yielding of steel sheets containing slits. *Journal of the Mechanics and Physics of Solids, 8*(2), 100–104.

Fan, T. G., & Zhang, G. Q. (2014). Laboratory investigation of hydraulic fracture networks in formations with continuous orthogonal fractures. *Energy, 74*, 164–173.

Griffith, A. A. (1920). The phenomena of rupture and flow in solids. *Philosophical Transactions of the Royal Society A Mathematical Physical and Engineering Sciences A, 221*(4), 163–198.

Griffith, A. A. (1921). The phenomena of rupture and flow in solids. *Philosophical Transactions of the Royal Society of London Series A, 221*(582–593), 163–198.

Gutierrez, M., & Youn, D. J. (2015). Effects of fracture distribution and length scale on the equivalent continuum elastic compliance of fractured rock masses. *Journal of Rock Mechanics and Geotechnical Engineering, 7*(6), 626–637.

Hamid, N. H., Anuar, S. A., Awang, H., & Kori, M. E. (2018). Experimental study on seismic behavior of repaired tunnel form building under cyclic loading. *Asian Journal of Civil Engineering, 19*, 343–354.

Hu, W., Ha, Y. D., & Bobaru, F. (2012). Peridynamic model for dynamic fracture in unidirectional fiber-reinforced composites. *Computer Methods in Applied Mechanics and Engineering, 217–220*, 247–261.

Irwin, G. R., Kies, J. A., & Smith, H. L. (1958). Fracture strength relative to onset and arrest of crack propagation. *Proc ASTM, 58*, 640–657

Jeffrey, R. G., Bunger, A. P., Lecampion, B., Zhang, X., Chen, Z. R., van As, A. & Mainguy, M. (2009). Measuring hydraulic fracture growth in naturally fractured rock. In *SPE Annual Technical Conference and Exhibition*, SPE-124919-MS.

Ji, Y., Zhuang, L., Wu, W., Hofmann, H., Zang, A., & Zimmermann, G. (2021). Cyclic water injection potentially mitigates seismic risks by promoting slow and stable slip of a natural fracture in granite. *Rock Mechanics and Rock Engineering*, 1–17.

Lamont, N., & Jessen, F. W. (1963). The effects of existing fractures in rocks on the extension of hydraulic fractures. *Journal of Petroleum Technology, 15*(2), 203–209.

Li, X., Xu, M., Wang, Y., Wang, G., Huang, J., Yin, W., & Yan, G. (2021). Numerical study on crack propagation of rock mass using the time sequence controlled and notched blasting method. *European Journal of Environmental and Civil Engineering, 26*(13), 6714–6732.

Li, S., Chen, Z., Li, W., Yan, T., Bi, F., & Tong, Y. (2023). An FE simulation of the fracture characteristics of blunt rock indenter under static and harmonic dynamic loadings using cohesive elements. *Rock Mechanics and Rock Engineering, 56*(4), 2935–2947.

Meglis, I. L., Chow, T., Martin, C. D., & Young, R. P. (2005). Assessing in situ microcrack damage using ultrasonic velocity tomography. *International Journal of Rock Mechanics and Mining Sciences, 42*(1), 25–34.

Miehe, C., & Mauthe, S. (2016). Phase field modeling of fracture in multi-physics problems. Part III.Crack driving forces in hydro-poro-elasticity and hydraulic fracturing of fluid-saturated porous media. *Computer Methods in Applied Mechanics Engineering, 304*, 619–655.

Miehe, C., Hofacker, M., & Welschinger, F. (2010). A phase field model for rate-independent crack propagation: Robust algorithmic implementation based on operator splits. *Computer Methods in Applied Mechanics and Engineering, 199*(45–48), 2765–2778.

Palaniswamy, K. K. W. G., & Knauss, W. G. (1978). On the problem of crack extension in brittle solids under general loading. *Mechanics Today*, 87–148.

Peng, P., Ju, Y., Wang, Y., Wang, S., & Gao, F. (2017). Numerical analysis of the effect of natural microcracks on the supercritical CO_2 fracturing crack network of shale rock based on bonded particle models. *International Journal for Numerical and Analytical Methods in Geomechanics, 41*(18), 1992–2013.

Ren, H., Zhuang, X., & Rabczuk, T. (2016). A new peridynamic formulation with shear deformation for elastic solid. *Journal of Micromechanics and Molecular Physics, 1*(2), 1650009.

Shin, H., & Santamarina, J. C. (2019). An implicit joint-continuum model for the hydro-mechanical analysis of fractured rock masses. *International Journal of Rock Mechanics and Mining Sciences, 119*, 140–148.

Su, H., Zhu, Z., & Wang, L. (2023). Experimental and numerical analysis of dynamic splitting mechanical properties and crack propagation for frozen sandstone. *International Journal of Rock Mechanics and Mining Sciences, 163*, 1365–1609.

Vahab, M., Hirmand, M. R., Jafari, A., & Khalili, N. (2021). Numerical analysis of multiple hydrofracture growth in layered media based on a non-differentiable energy minimization approach. *Engineering Fracture Mechanics, 241*, 107361.

Wang, Y., & Zhang, X. (2022). Dual bilinear cohesive zone model-based fluid-driven propagation of multiscale tensile and shear fractures in tight reservoir. *Engineering Computations, 39*(10), 3416–3441.

Wang, Y., Lin, H., Zhao, Y., Li, X., Guo, P., & Liu, Y. (2020). Analysis of fracturing characteristics of unconfined rock plate under edge-on impact loading. *European Journal of Environmental and Civil Engineering, 24*(14), 2453–2468.

Wang, Y., Huang, J., & Wang, G. (2022a). Numerical analysis for mining-induced stress and plastic evolution involving influencing factors: High in situ stress, excavation rate and multilayered heterogeneity. *Engineering Computations, 39*(8), 2928–2957.

Wang, Y., Wang, J., & Li, L. (2022b). Dynamic propagation behaviors of hydraulic fracture networks considering hydro-mechanical coupling effects in tight oil and gas reservoirs: A multi-thread parallel computation method. *Computers and Geotechnics, 152*, Article 105016.

Wang, Y., Feng, R., Li, D., & Peng, R. (2023). Numerical analysis for tunnelling-induced stress and plastic evolution causing instability of multilayered surrounding rock by varying three-dimensional in situ stresses. *Engineering Computations*. https://doi.org/10.1108/EC-12-2022-0715.

Warpinski, N. R., & Teufel, L. W. (1987). Influence of geologic discontinuities on hydraulic fracture propagation. *Journal of Petroleum Technology, 39*(2), 209–220.

Yang, T. H., Tham, L. G., Tang, C. A., Liang, Z. Z., & Tsui, Y. (2004). Influence of heterogeneity of mechanical properties on hydraulic fracturing in permeable rocks. *Rock Mechanics and Rock Engineering, 37*(4), 251–275.

Zhu, W. C., & Tang, C. A. (2006). Numerical simulation of Brazilian disk rock failure under static and dynamic loading. *International Journal of Rock Mechanics and Mining Sciences, 43*(2), 236–252.

Chapter 5
Center- and Edge-Type Intersections of Hydraulic Fracture Network Under Varying Crossed Natural Fractures and Fluid Injection Rate

5.1 Introduction

Hydrofracturing operations in naturally fractured reservoirs demonstrate that natural fracture properties (including orientation, spacing, length, and persistence) fundamentally govern hydraulic fracture network intersections as shown in Fig. 5.1 (Cheng et al., 2015; Fatahi et al., 2017; Rahman & Rahman, 2013; Xiong & Ma, 2022). These intersections generate complex fracture networks through interactions between natural and hydraulic fractures (Dehghan et al., 2015; Ghaderi et al., 2018; Hou et al., 2016; Shakib et al., 2015; Taleghani et al., 2016; Wasantha et al., 2017). Fluid injection rates critically influence propagation dynamics, where elevated rates increase fracture velocity and alter mechanical behaviors at natural fracture interfaces. High-rate injection induces rapid reservoir pressurization, concentrating energy near fracture tips to drive rock failure while curtailing matrix leak-off (Suo et al., 2020; Yan & Yu, 2022). The resultant fracture network morphologies vary significantly across natural fracture configurations and injection parameters, complicating the evaluation and optimization of gas recovery in tight unconventional reservoirs.

Current research examines how natural fracture properties (orientation, spacing, length, and persistence) govern hydraulic fracture propagation and intersection behaviors during stimulation. Fracture orientation defines approach angles between hydraulic and natural fractures, critically governing final network morphology (Sanchez et al., 2020; Zhao et al., 2022). Acute approach angles facilitate natural fracture activation, causing hydraulic fractures to deflect along natural fractures while activating weak interfaces that subsequently propagate along maximum in-situ stress directions (Zhou et al., 2017). Reduced approach angles and stress differentials enhance hydraulic-natural fracture connectivity and continued propagation (Li et al., 2020; Song et al., 2020), whereas 45° angles optimize natural fracture activation and

Y. Wang, *Deflection of Fracturing Fracture Network Disturbed by Discontinuity and Multiple Fractures in Rock*, https://doi.org/10.1007/978-981-95-1569-1_5

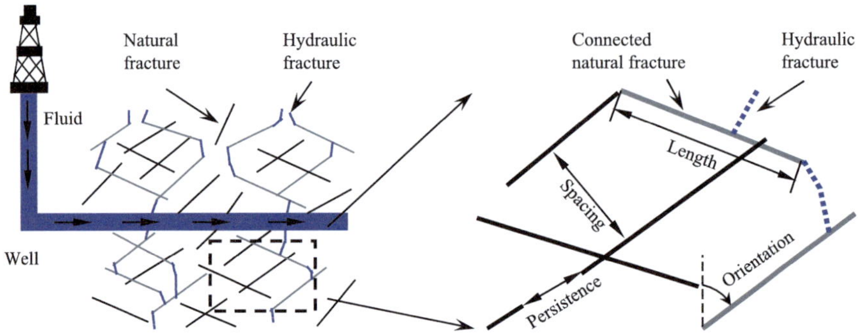

Fig. 5.1 Schematic simulation of dynamic propagation and intersection of hydraulic fractures and pre-existing crossed natural fractures

maximize complexity (Guo et al., 2015; Wang et al., 2018a, 2018b). Natural fracture density controlled by spacing, length, and persistence directs hydraulic fracture trajectories and connected fracture areas, with propagation preferentially following dominant dense networks to establish interconnected systems (Cao et al., 2022; Wang et al., 2016).

Hydraulic fractures enhance natural fracture connectivity and expand fracture network surface area, creating efficient recovery pathways for unconventional hydrocarbons (Wang et al., 2019). Correlations exist between natural fracture parameters and resultant network morphologies, where optimized flow channels can be engineered through morphological control (Li, 2022; Makedonska et al., 2020). Fundamental relationships remain unclear among natural fractures, stimulation parameters (e.g., injection rates), fracture architectures, and production outcomes. Optimizing fracturing efficacy thus requires quantitative evaluation of morphology-production relationships based on fracture sensitivity factors and injection parameters. Whereas our prior research characterized center- and edge-type intersections with large-scale natural fractures (Wang et al., 2021), elucidating mechanisms linking fracture properties, morphologies, and production, these findings may not extend to small-scale systems. Consequently, this investigation establishes small-scale fracture networks to analyze intersection behaviors and morphology-production relationships under variable natural fracture configurations and injection parameters.

The remaining parts of this chapter are as follows: Sect. 5.2 introduces the combined finite element-discrete element method and model considering hydro-mechanical coupling. Section 5.3 introduces the numerical models of fractured reservoir embedded discrete fracture networks, as well as the typical cases that consider the sensitivity factors including the orientation, spacing, length, persistence of pre-existing crossed natural fractures and fluid injection rate. Section 5.4 introduces the results and discussion for sensitivity factors of pre-existing natural fractures, quantitative length of fracture networks, and gas production in fractured reservoirs. Section 5.5 summarizes the main conclusions of this study.

5.2 Combined Finite Element-Discrete Element Method and Model Considering Hydro-Mechanical Coupling

5.2.1 Governing Partial Differential Equations

The mechanical governing equation for the solid stress field is:

$$\boldsymbol{L}^T(\boldsymbol{\sigma}^e - \alpha \boldsymbol{m} p_s) + \rho_b \boldsymbol{g} = 0 \tag{5.1}$$

where $\boldsymbol{L}$ is the spatial differential operator; $\boldsymbol{\sigma}^e$ is the effective stress tensor; α is the effective stress coefficient, also known as Biot's coefficient, and its value is between 0 and 1, which is used to characterize the effect of pore pressure on rock deformation; $\boldsymbol{m}$ is the identity tensor; p_s is the pore fluid pressure; ρ_b is the wet bulk density, and $\boldsymbol{g}$ is the gravity vector.

The governing equations for liquid seepage and fracture fluid flow and the fluid network equation for fluid flow in the fracture region combine mass conservation along with Darcy's law and are given as follows:

$$\mathrm{div}\left[\frac{k}{\mu_l}(\nabla p_l - \rho_l g)\right] = \left(\frac{\phi}{K_l} + \frac{\alpha - \phi}{K_s}\right)\frac{\partial p_l}{\partial t} + \alpha \frac{\partial \varepsilon_v}{\partial t} \tag{5.2}$$

$$\frac{\partial}{\partial x}\left[\frac{k^{fr}}{\mu_n}\left(\nabla p_n - \rho_{fn}\boldsymbol{g}\right)\right] = S^{fr}\frac{\mathrm{d}p_n}{\mathrm{d}t} + \alpha(\Delta \dot{e}_\varepsilon) \tag{5.3}$$

where k is the intrinsic permeability of the porous media; μ_l is the viscosity of the pore liquid; p_l is the pore liquid pressure; ρ_l is the density of the pore liquid; ϕ is the porosity of the porous media; K_l is the bulk stiffness of the pore liquid; K_s is the bulk stiffness of the solid grains; ε_v is the volumetric strain of the porous media; k^{fr} is the intrinsic permeability of the fractured region; μ_n is the viscosity of the fracturing fluid; p_n is the fracturing fluid pressure; ρ_{fn} is the density of the fracture fluid; S^{fr} is the storage coefficient (which is effectively a measure of the compressibility of the fractured region when a fluid is present); and $\Delta \dot{e}_\varepsilon$ is the aperture strain rate.

The governing equations of gas seepage and gas network for gas recovery and production combine mass conservation along with Darcy's law and are given as follows:

$$div\left[\frac{k(p_g)}{\mu_g}\nabla p_g - \rho_g g\right] = \left[\phi\frac{\partial \rho_g}{\partial p_g} + (\rho_g - q)\frac{\partial \phi}{\partial p_g} + (1-\phi)\frac{\partial q}{\partial p_g}\right]\frac{\partial p_g}{\partial t} \tag{5.4}$$

$$\frac{\partial}{\partial x}\left[\frac{K^{fr}}{\mu_g}(\nabla p_g - \rho_g \boldsymbol{g})\right] = \phi\left(C_g - \frac{\rho_g}{Z}\frac{\partial \rho_g}{\partial Z}\right)\frac{\partial \rho_g}{\partial t} \tag{5.5}$$

where μ_g is the viscosity of the pore gas; p_g is the pore gas pressure; ρ_g is the density of the pore gas; ϕ is the porosity of the porous media; q is the mass of adsorbed gas per unit volume; C_g is the gas compressibility; Z is the gas compressibility factor. The discretization scheme for above by governing equations can be found in literatures (Wang et al., 2021).

5.2.2 Discrete Fracture Network Model

In two-dimensional (2D) discrete fracture network (DFN), a fracture is represented as a line segment connecting two points in the region:

$$H \equiv [P, Q] \tag{5.6}$$

where $P = (x_1, y_1)$, and $Q = (x_2, y_2)$ are the endpoints of H. We can also represent a fracture using four parameters, $w = (w_1, w_2, w_3, w_4)$, where (w_1, w_2) are the coordinates of the centre and w_3 and w_4 are, respectively, the orientation and the length of the fracture (Seifollahi et al., 2014; Wang et al., 2021). DFN model gives more practical consideration to the fracture occurrence by fully discretizing the matrix and fracture system in the modelling process, and can more accurately describe the flow law of reservoir under heterogeneous conditions. As shown in Fig. 5.2, the fractures are discretized by finite element method, and the matrix is discretized by triangular elements with the fracture elements as the grid constraint (Karimi-Fard and Firoozabadi, 2001).

In this method, the fractures are reduced and homogenized. The discrete DFN model can be applied to any complex fracture structure theoretically because of its flexibility in division.

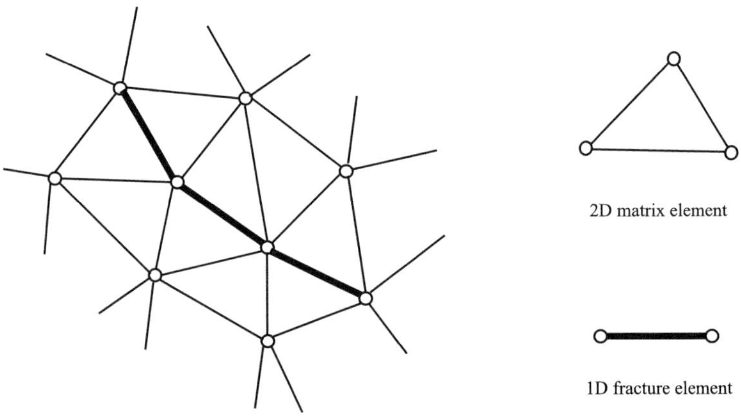

Fig. 5.2 Discretization method of discrete fractured media

$$\int_{\Omega} FEQ\mathrm{d}\Omega = \int_{\Omega_m} FEQ\mathrm{d}\Omega_m + e \times \int_{\overline{\Omega}_f} FEQ\mathrm{d}\overline{\Omega}_f \tag{5.7}$$

where FEQ represent the flow equations of the matrix and the fracture media; Ω represents the whole domain; Ω_m represents the matrix; $\overline{\Omega}_f$ represents the one-dimensional (1D) fracture part of the domain; and the fracture width is represented by e, which is the coefficients before one dimensional integration

5.3 Numerical Models of Fractured Reservoir Embedded Discrete Fracture Networks

5.3.1 Geometrical Models

This investigation employs a 50 m × 30 m two-dimensional geometrical model featuring two discrete fracture network (DFN) sets representing pre-existing natural fractures (Fig. 5.3). A centrally positioned horizontal well contains an initial 2 m perforation cluster, while pre-existing fractures occupy a distinct 20 m × 10 m subdomain. To enhance computational accuracy, Fig. 5.4 illustrates the hydrofracturing finite element model implementing localized mesh refinement around the fracture domain. Material properties governing the numerical simulation are cataloged in Table 5.1.

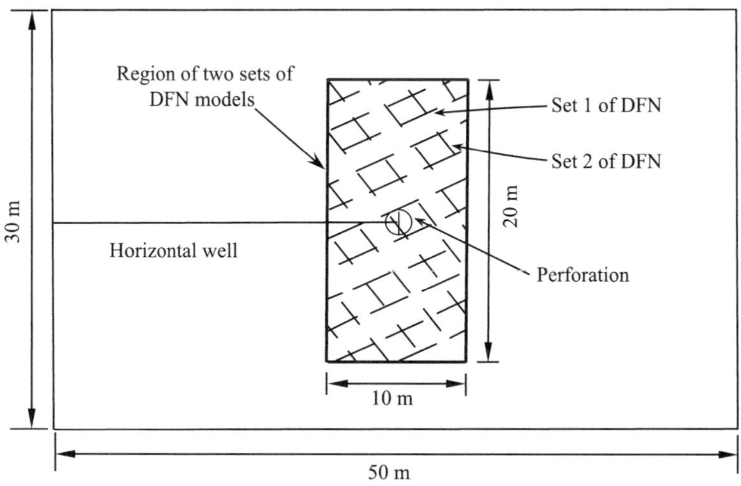

Fig. 5.3 Geometrical model of hydrofracturing in fractured reservoir with two sets of DFN models

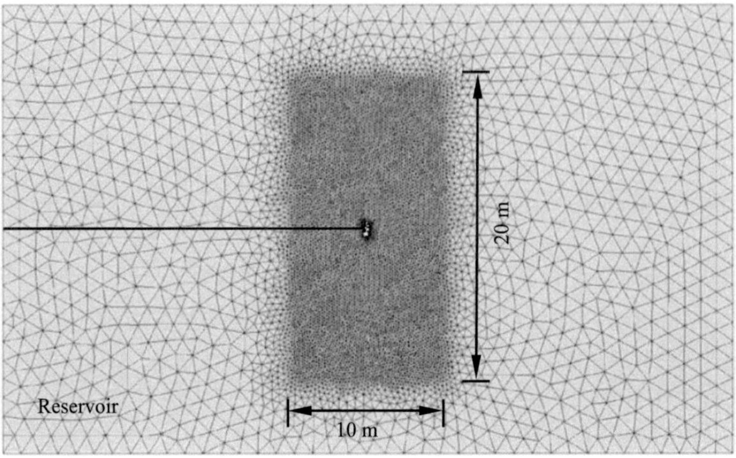

Fig. 5.4 Finite element model of hydrofracturing in fractured reservoir

Table 5.1 Physical parameters of numerical model of hydrofracturing

Parameters	Value
Depth of horizontal well/m	4000
Horizontal in-situ stress in x direction S_h/MPa	40
Vertical in-situ stress in y direction S_v/MPa	44
Pore pressure p_s/MPa	30
Biot's coefficient α	0.8
Young's modulus E/GPa	31
Poisson's ratio ν	0.2
Porosity ϕ	0.05
Tensile strength σ_t/MPa	1.0
Fracture energy G_f/(N · m)	50
Leak-off coefficient $C_I/(m^3/s^{1/2})$	0.1×10^{-6}
Leak-off coefficient $C_{II}/(m^3/s^{1/2})$	0.1×10^{-6}
Density ρ_b/(kg/m^3)	2.615×10^3
Permeability k/(nD)	50
Gravity g/(m/s^2)	9.81
Liquid density of the pore fluid ρ_g/(kg/m^3)	1×10^3
Liquid density of the fracturing fluid ρ_{fn}/(kg/m^3)	1×10^3

5.3.2 Cases Study for Typical Pre-existing Crossed Natural Fractures

Table 5.2 details DFN configurations and injection parameters, with sensitivity factors incremented as follows: orientation (30° intervals), spacing/length/persistence (0.5 m increments), and fluid injection rate (0.05 m^3/s increments). Sixteen numerical cases organized into five parametric groups systematically evaluate how these factors govern hydraulic fracture propagation. Case I serves as the benchmark for comparative analysis across all scenarios.

(a) Cases I, II, III, and IV evaluate orientation effects on hydraulic fracture propagation;
(b) Cases I, V, VI, and VII assess spacing influences;
(c) Cases I, VIII, IX, and X quantify length dependencies;
(d) Cases I, XI, XII, and XIII examine persistence impacts;
(e) Cases I, XIV, XVI, and XVI analyze injection rate consequences.

Figure 5.5 illustrates the morphology of natural fractures in the aforementioned cases. The natural fracture network is formed by two sets of DFN natural fractures, thereby enabling observation of differences in propagation and intersection behaviors when hydraulic fractures encounter these pre-existing discontinuities. Computational analyses for these cases were performed using the ELFEN TGR program package (Rockfield Software Ltd., 2016), with simulations executed on an Intel® Core™ 3.40 GHz processor.

5.4 Results and Discussion

5.4.1 Sensitivity Factors of Pre-existing Natural Fractures

5.4.1.1 Orientation of Inclined Natural Fractures

The results of the orientation of the initial pre-existing inclined natural fractures for the sensitivity analyses are shown in Fig. 5.6. The orientation angles of the two sets of DFN natural fractures are 30, 60°, 90°, 120° and 105°, 135°, 165°, 195°, respectively. There are two types of fracture network morphologies when hydraulic fractures intersect with natural fractures:

(a) Edge-type propagation is the result of the intersection of hydraulic fractures and the edge of the natural fractures, as shown in Fig. 5.6a and b. The hydraulic fracture can intersect with the edge of the natural fracture and lead to edge-type propagation, which is conducive for the fracture propagating towards the area farther away from the perforation. In edge-type propagation, when the approach

Table 5.2 Pre-existing fracture set of DFN models and injection rate

State		Cases															
		I	II	III	IV	V	VI	VII	VIII	IX	X	XI	XII	XIII	XIV	XV	XVI
DFN Set 1	Orientation/(°)	60	30	90	120	60	60	60	60	60	60	60	60	60	60	60	60
	Spacing/m	1.5	1.5	1.5	1.5	1	2	2.5	1.5	1.5	1.5	1.5	1.5	1.5	1.5	1.5	1.5
	Length/m	2.5	2.5	2.5	2.5	2.5	2.5	2.5	2	3	3.5	2.5	2.5	2.5	2.5	2.5	2.5
	Persistence/m	1	1	1	1	1	1	1	1	1	1	0.5	1.5	2	1	1	1
DFN Set 2	Orientation/(°)	135	105	165	195	135	135	135	135	135	135	135	135	135	135	135	135
	Spacing/m	1.25	1.25	1.25	1.25	0.75	1.75	2.25	1.25	1.25	1.25	1.25	1.25	1.25	1.25	1.25	1.25
	Length/m	1.75	1.75	1.75	1.75	1.75	1.75	1.75	1.25	2.25	2.75	1.75	1.75	1.75	1.75	1.75	1.75
	Persistence/m	1.25	1.25	1.25	1.25	1.25	1.25	1.25	1.25	1.25	1.25	0.75	1.75	2.25	1.25	1.25	1.25
Injection rate/(m^3/s)		0.075	0.075	0.075	0.075	0.075	0.075	0.075	0.075	0.075	0.075	0.075	0.075	0.075	0.025	0.125	0.175

angle between hydraulic fractures and natural fractures is small enough, the hydraulic fractures will be reoriented and activate the natural fractures.

(b) Center-type propagation is the result of the intersection of hydraulic fractures and crossed clusters of natural fractures, as shown in Fig. 5.6c and d. The hydraulic fracture may intersect with the natural fracture cluster to form a center-type propagation.

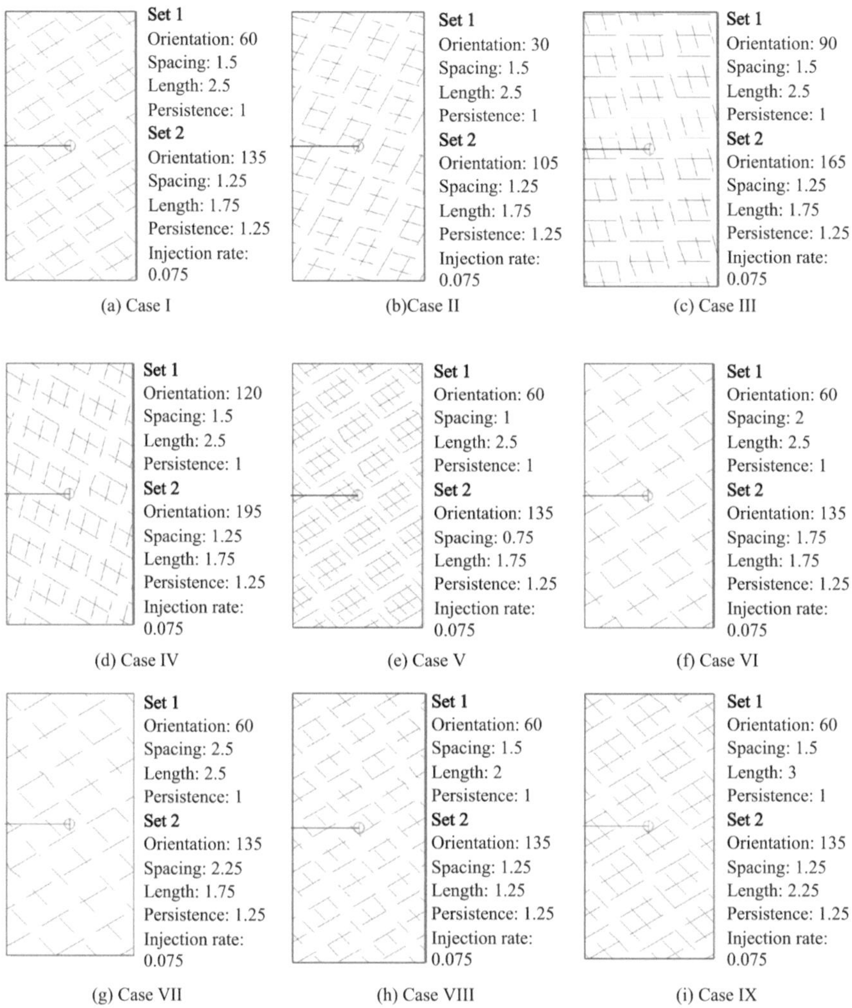

Fig. 5.5 Cases for the orientation, spacing, length, persistence of crossed natural fractures and fluid injection rate

5.4.1.2 Spacing of Adjacent Natural Fractures

As shown in Fig. 5.7a and b, when the fracture spacing is small, the hydraulic fracture is more likely to propagate from the initial perforation to the intersection with the natural fracture cluster, and the center-type propagation occurs. As shown in Fig. 5.7c, With the increase of the fracture spacing and the decrease of the natural fracture density, the hydraulic fracture is more likely to propagate to form the edge-type propagation. The increasing spacing will cause hydraulic fractures to propagate along a straight line (maximum principal stress direction), forming a single fracture or intersect with the edge of natural fractures.

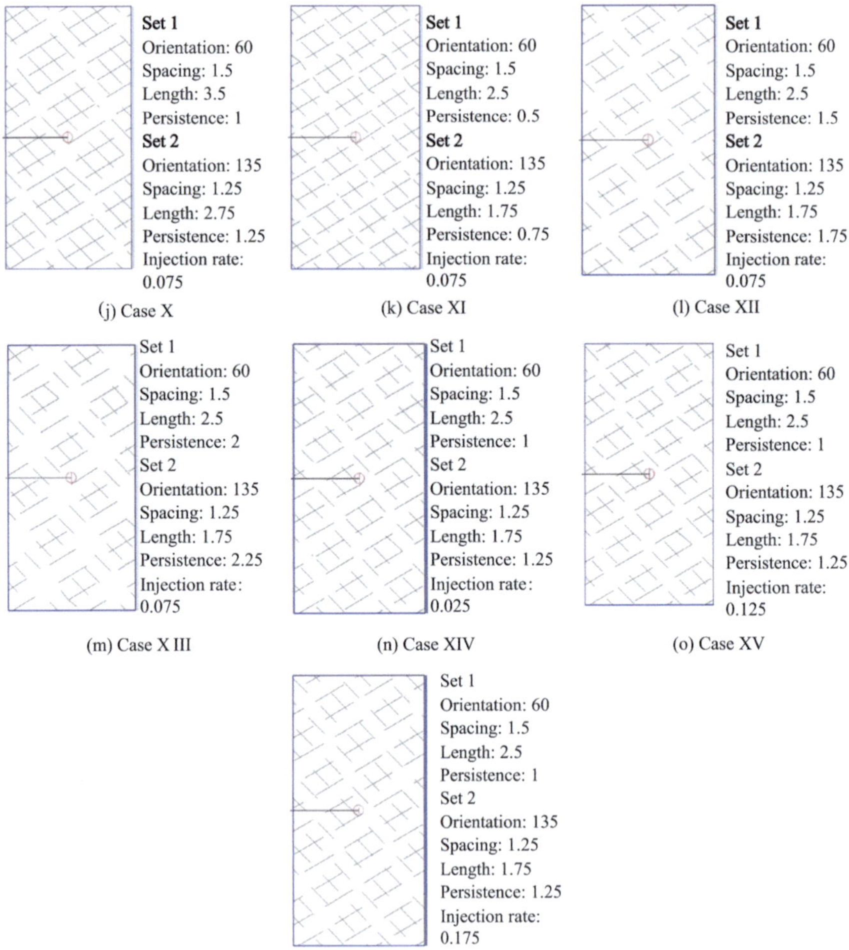

(j) Case X (k) Case XI (l) Case XII

(m) Case X III (n) Case XIV (o) Case XV

(p) Case XVI

Fig. 5.5 (continued)

5.4.1.3 Length of Straight Natural Fractures

The results of stress, dynamic propagation, and intersection fractures for length of straight natural fractures are shown in Fig. 5.8. As shown in Fig. 5.8a, when the length of straight natural fractures is small, the hydraulic fractures may tendentiously propagate along the edge of the natural fracture, and propagate in the rock matrix to form some single fractures. As shown in Fig. 5.8b and c, the long length of straight natural fractures will make natural fractures interconnected, and the hydraulic fractures is prone to intersect and activate the natural fractures; once the hydraulic

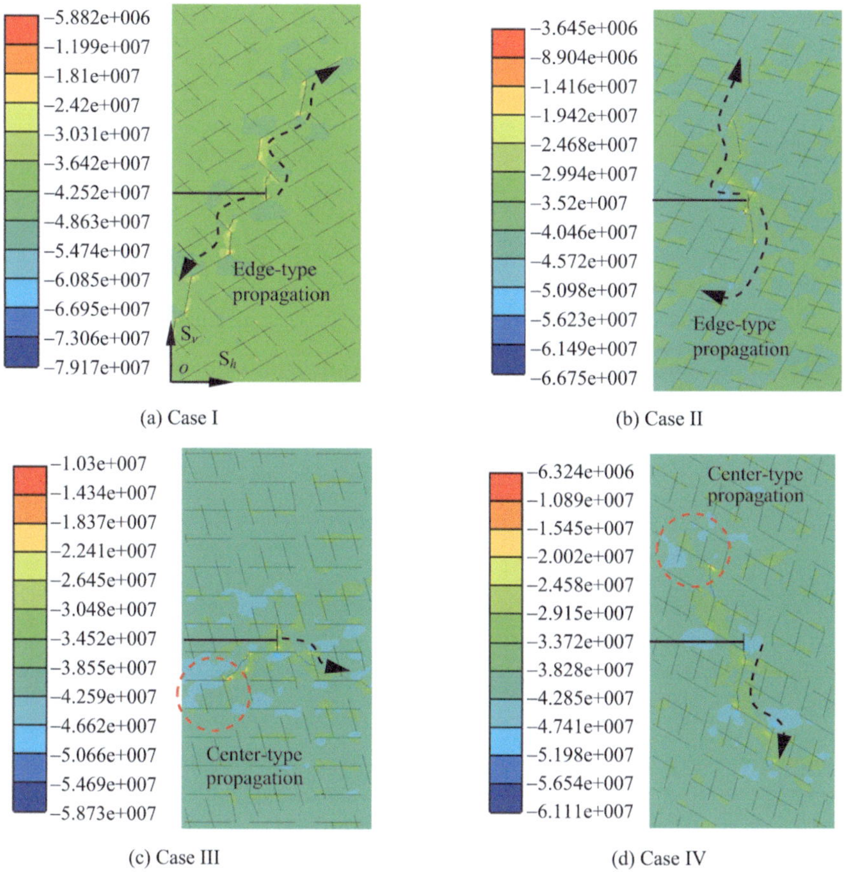

Fig. 5.6 Results of stress (first principal stress, Pa), dynamic propagation, and intersection fractures for orientation of inclined natural fractures

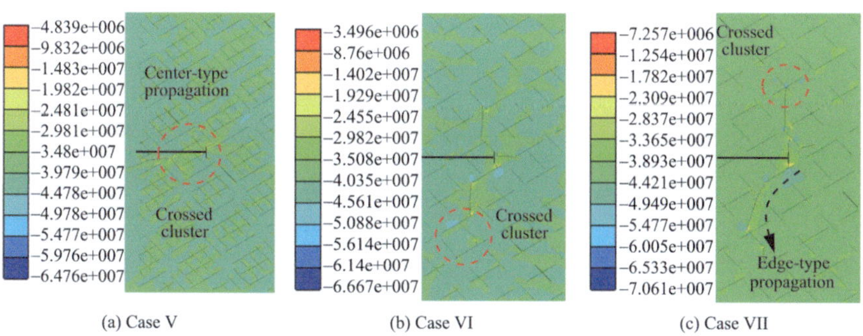

Fig. 5.7 Results of stress (first principal stress, Pa), dynamic propagation, and intersection fractures for spacing of adjacent natural fractures

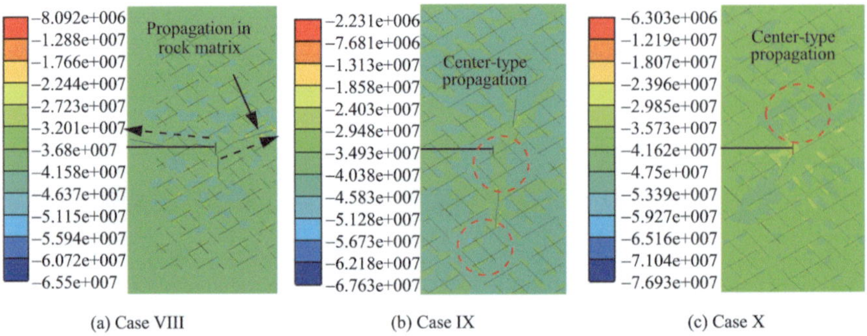

Fig. 5.8 Results of stress (first principal stress, Pa), dynamic propagation, and intersection fractures for length of straight natural fractures

fracture intersects with the original fracture network, it is conducive to form the center-type propagation of fractures.

5.4.1.4 Persistence of Adjacent Natural Fractures

As shown in Fig. 5.9a, a small persistence of adjacent natural fractures may cause the tips of the natural fracture to be relatively close; once the hydraulic fracture intersects with the natural fracture, the hydraulic fracture will deflect and be prone to form center-type propagation, and this center-type fracture network will facilitate the recovery of oil and gas in the reservoir. As shown in Fig. 5.9b and c, with the increase of persistence of adjacent natural fractures, the hydraulic fractures mostly intersect with the edge of the natural fractures to form the edge-type propagation; due to the larger spacing between natural fracture clusters, hydraulic fractures will intersect and propagate with natural fractures, making it easier to form a relatively simple morphology of fracture network.

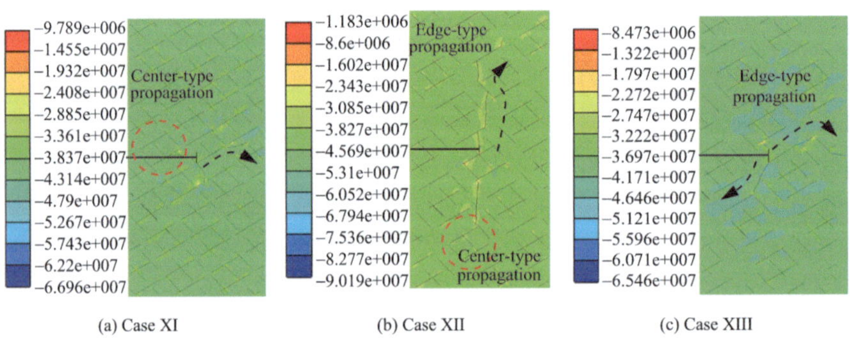

Fig. 5.9 Results of stress (first principal stress, Pa), dynamic propagation, and intersection fractures for persistence of adjacent natural fractures

5.4.1.5 Fluid Injection Rate

The low injection rate (Fig. 5.10a, 0.025 m^3/s of fluid injection rate) drives hydraulic fractures propagation, making it easier to encounter and connect with natural fractures in the surrounding domain around the fracture tips; at this stage, the fractures exhibit quasi-static or steady-state propagation, and the fracture propagation speed is slow. The high injection rate (Fig. 5.10b and c, 0.125 and 0.175 m^3/s of fluid injection rate) drives the rapid propagation of hydraulic fractures, making it easier for them to propagate outward and form a single and long fracture during the fracture propagation process; at this stage, the fractures exhibit quasi dynamic propagation and the fracture propagation speed is relatively fast.

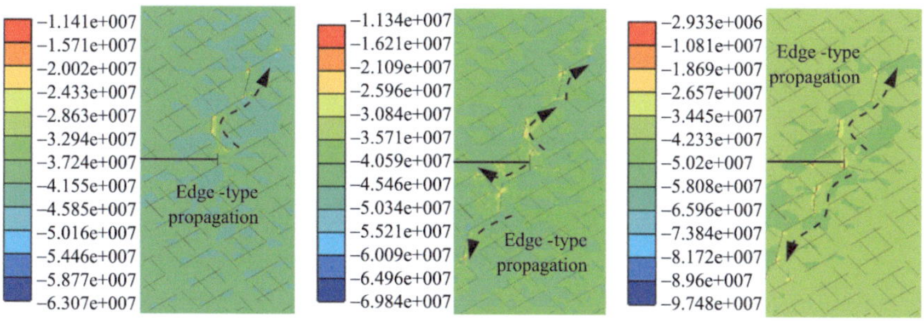

(a) Case XIV (b) Case XV (c) Case XVI

Fig. 5.10 Results of stress (first principal stress, Pa), dynamic propagation, and intersection fractures for fluid injection rate

5.4.2 *Quantitative Length of Fracture Networks*

To investigate the influences of the sensitivity factors of natural fractures and injection rate on the fracture network, the quantitative results of the length of the fracture networks may be derived and discussed in detail as follows. Figure 5.11 shows the evolution of fracture length involving the influence of the orientation, spacing, length, persistence of crossed natural fractures and fluid injection rate. Table 5.3 lists the final lengths of hydraulic fractures and connected natural fractures in each case, and the results with longer fracture lengths in each comparative group are highlighted with underlines; the results are also shown in the Fig. 5.12 for a more intuitive comparison and analysis:

(a) The orientation variation of natural fracture determines the intersection and propagation direction of hydraulic fractures (Fig. 5.11a): when the orientation angles of natural fractures are 90° and 165° (Case III), hydraulic fractures intersect with natural fracture clusters, forming center-type intersection, arising the longest fracture length; once the orientation angles of natural fractures are 120°

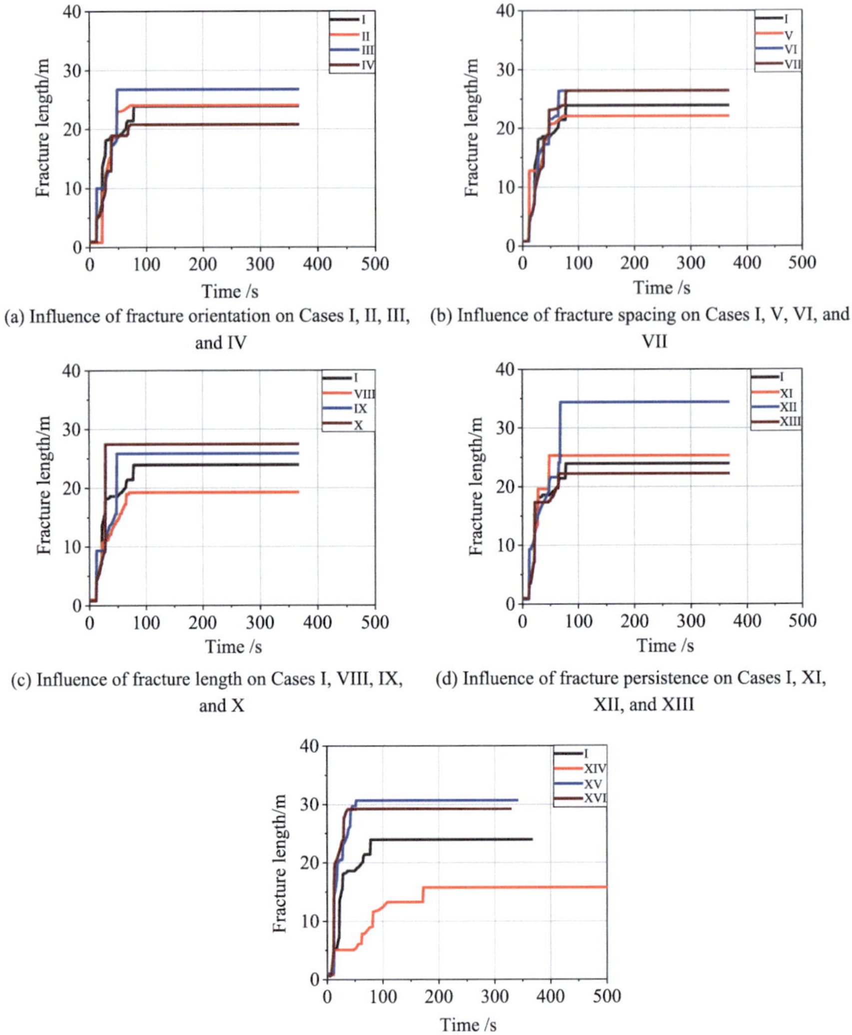

(a) Influence of fracture orientation on Cases I, II, III, and IV

(b) Influence of fracture spacing on Cases I, V, VI, and VII

(c) Influence of fracture length on Cases I, VIII, IX, and X

(d) Influence of fracture persistence on Cases I, XI, XII, and XIII

(e) Influence of injection rate on Cases I, XIV, XV, and XVI

Fig. 5.11 Evolution for fracture length involving the influence of the orientation, spacing, length, persistence of crossed natural fractures and fluid injection rate

and 195° (Case IV), most hydraulic fractures propagate along the edge of natural fractures and form edge-type propagation, and few natural fracture clusters are activated by hydraulic fractures to form the shortest fracture length.

(b) Fig. 5.11b shows the fracture length considering the sensitivity factor of the spacing of natural fractures: in Case V, when the spacing between natural fractures is narrow (1 and 0.75 m for the two sets of natural fractures, respectively),

Table 5.3 Final length of hydraulic fractures and connected natural fractures

Sensitivity factor	Cases	Length of hydraulic fracture/m	Length of hydraulic and connected natural fractures/m	Initial natural fractures/m
Orientation	I	9.24	23.93	186
	II	7.59	24.09	187
	III	5.53	26.78	196
	IV	5.02	20.77	185
Spacing	V	5.86	22.11	294
	VI	8.68	26.43	137
	VII	11.18	26.43	108
Length	VIII	10.74	19.24	167
	IX	6.45	25.83	202
	X	5.69	27.44	211
Persistence	XI	5.28	25.28	222
	XII	10.37	34.34	164
	XIII	6.98	22.23	142
Injection rate	XIV	5.31	15.76	185
	XV	10.19	30.69	183
	XVI	12.94	29.19	186

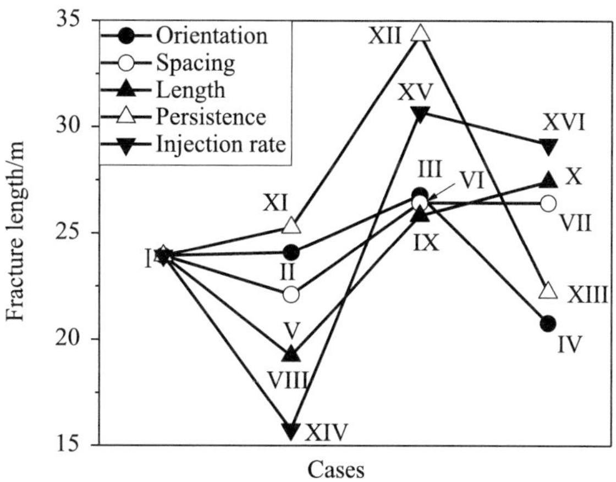

Fig. 5.12 Comparison of fracture length in different cases

hydraulic fractures are most likely to activate and connect the natural fracture clusters, some local and short fracture networks are formed.

(c) Fig. 5.11c shows the fracture length considering the sensitivity factor of the length of natural fractures: in Case X, when the length of natural fractures is large (3.5 m and 2.75 m for two sets of natural fractures, respectively), hydraulic fractures are prone to connect the natural fracture clusters and form long fracture networks.

(d) Fig. 5.11d shows the fracture length considering the sensitivity factor of the persistence of natural fractures: in Case XI, when the persistence of natural fractures is relatively short (0.5 m and 0.75 m for two sets of natural fractures, respectively), hydraulic fractures are most likely to activate and connect natural fracture clusters, forming some long natural fractures. Due to center-type intersection of hydraulic fractures and natural fracture clusters, Case XII arises the longest fracture length.

(e) Fig. 5.11e shows the fracture length considering the sensitivity factor of the fluid injection rate of natural fractures: in Cases XV and XVI, when the injection rate is high (0.125 and 0.175 m^3/s of injection rate), the fracture length is significantly larger than the low injection rate in Case XIV (0.025 m^3/s of injection rate). This indicates that the high injection rate drives the rapid propagation of hydraulic fractures, and the fractures exhibit quasi dynamic propagation and the long fractures are induced.

5.4.3 Gas Production in Fractured Reservoirs

Figure 5.13 shows the representative results of gas pressure at the initial gas production stage, and other similar results are not listed further. The gas pressure near hydraulic fractures and connected natural fractures is lower than other areas in the overall reservoir, which forms the low-pressure seepage channels allowing gas recover and production. When small-scale natural fractures form small-scale and aggregated center- and edge-type intersections of fracture network, the increased fractures gather together to form the clustered low pressure area.

To investigate the influences of the sensitivity factors of natural fractures and injection rate on the gas production, the quantitative results of the gas production volume may be derived and discussed in detail as follows. Figure 5.14 shows the evolution

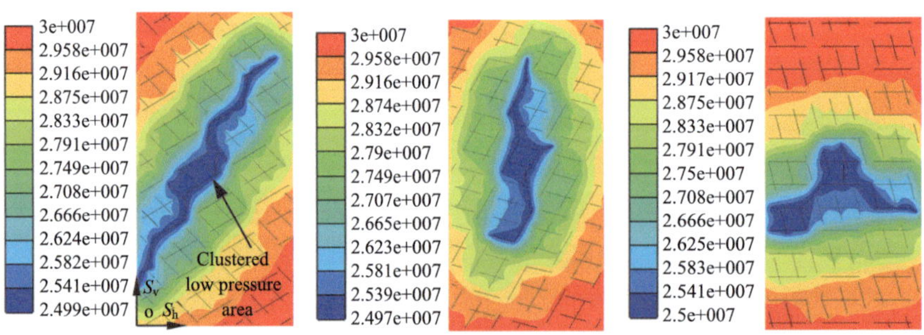

(a) Case I (b) Case II (c) Case III

Fig. 5.13 Representative results of gas pressure (Pa) at the initial gas production stage of gas production volume involving the influence of the orientation, spacing, length, persistence of crossed natural fractures and fluid injection rate. With the increase of time, the gas quantity of each case increases, and the curve slope of the gas production decreases, indicating that the gradient of gas production quantity increase gradually slows down over time.

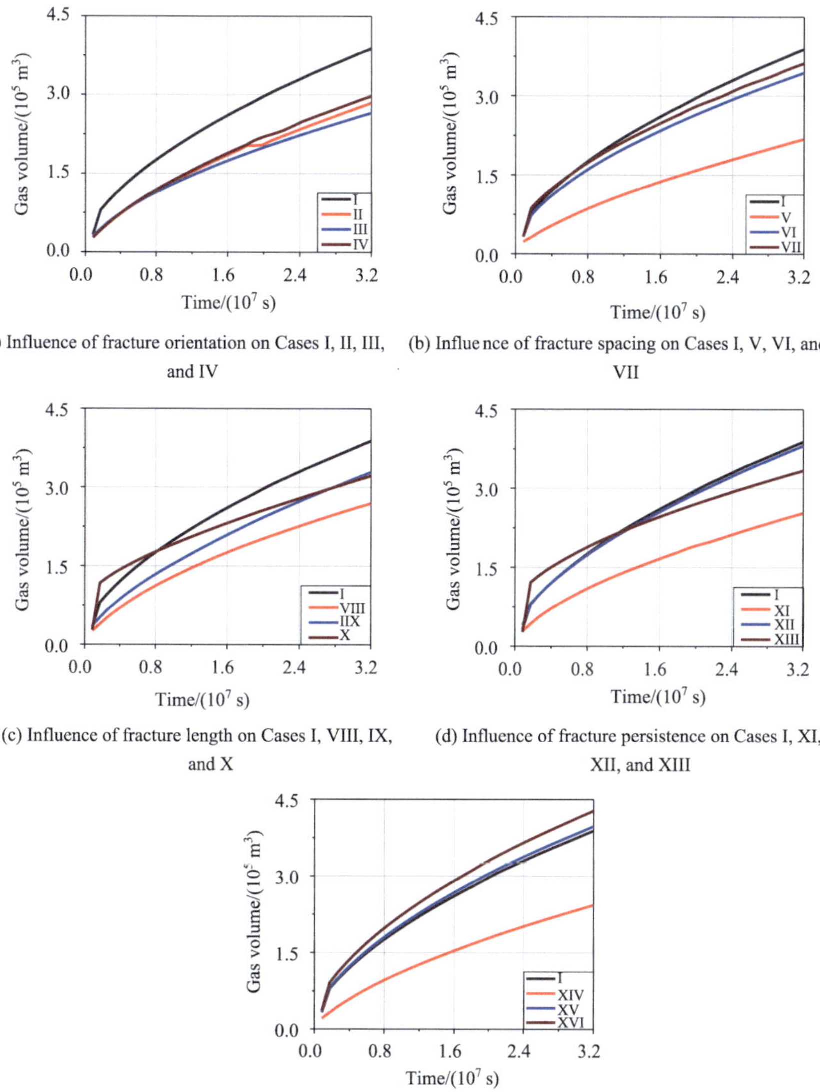

(a) Influence of fracture orientation on Cases I, II, III, and IV

(b) Influence of fracture spacing on Cases I, V, VI, and VII

(c) Influence of fracture length on Cases I, VIII, IX, and X

(d) Influence of fracture persistence on Cases I, XI, XII, and XIII

(e) Influence of injection rate on Cases I, XIV, XV, and XVI

Fig. 5.14 Evolution for gas production volume involving the influence of the orientation, spacing, length, persistence of crossed natural fractures and fluid injection rate

Table 5.4 lists the final length of hydraulic fractures and connected natural fractures and gas production volume, and the numerical cases were reordered according to the corresponding fracture length (the results with longer fracture lengths are also highlighted with underlines as Table 5.3); the results are also shown in the Fig. 5.15 for a more intuitive comparison and analysis. Through comparison, it can be seen that the length of fractures during the fracturing process is positively correlated with gas production. Some cases (Cases VI, VII, III, X, XVI, XV, and XII) can generate longer fractures, which maintain relatively high gas quantity throughout the entire long-term production process.

To quantitatively obtain the relationship, the fitting curve is derived via the least squares method, as shown in Fig. 5.16a, and the equation of the curve is as follows:

$$V = -141 \times L^2 + 16379 \times L \tag{5.8}$$

$$R^2 = 0.98 \tag{5.9}$$

where V is the volume of gas production (m^3); L is the fracture length (m); and R^2 is R-squared of the correlation coefficient.

Table 5.4 Final length of hydraulic fractures and connected natural fractures and gas production volume

Cases	Length of hydraulic and connected natural fractures/m	Gas production volume/(10^5 m^3)
XIV	15.76	2.43
VIII	19.24	2.69
IV	20.77	2.97
V	22.11	2.18
XIII	22.23	3.33
I	23.93	3.88
II	24.09	2.84
XI	25.28	2.53
IX	25.83	3.28
VI	26.43	3.44
VII	26.43	3.62
III	26.78	2.65
X	27.44	3.21
XVI	29.19	4.27
XV	30.69	3.97
XII	34.34	3.81

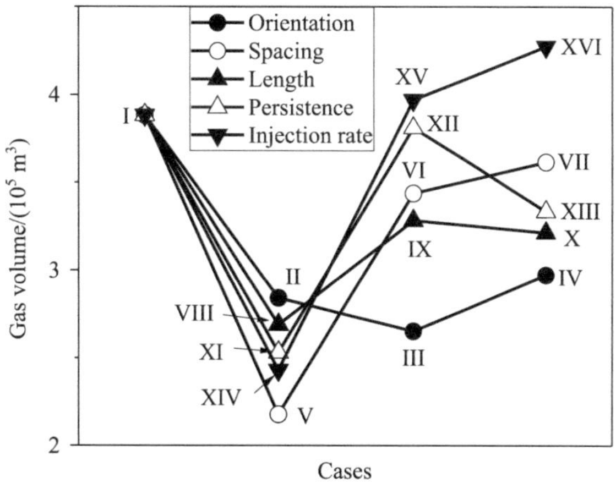

Fig. 5.15 Comparison of gas production volume in different case

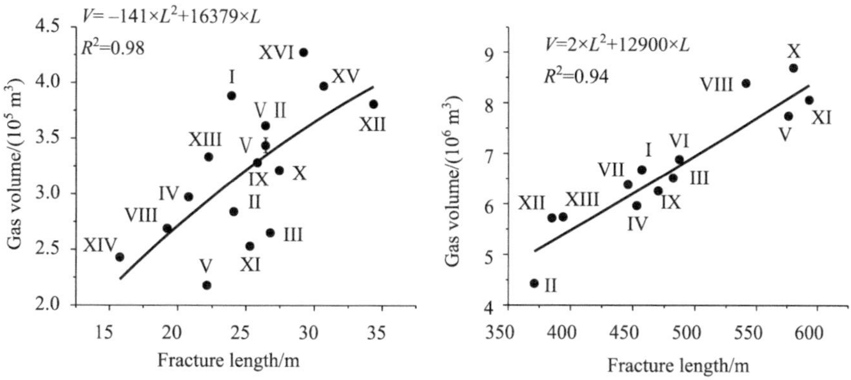

(a) Small-scale nature fracture

(b) Large-scale nature fracture (Wanget al., 2021)

Fig. 5.16 Relationship of fracture length and gas production volume

This equation indicates the relationship between the length of small-scale fractures and gas production, which is similar to the relationship between the length of large-scale fractures and gas production as shown in Fig. 5.16b (Wang et al., 2021), reflecting the correlation between the lengths of small- and large-scale fractures and gas production. Therefore, for the sensitivity factors (orientation, spacing, length, and persistence) of natural fractures and fluid injection rate, the formed center-type intersections of hydraulic fracture network may generate long fracture length, which is prone to improving gas production; when the hydrofracturing scheme is designed, it is crucial to actively promote the center-type intersections of hydraulic fracture network based on the morphology of natural fractures.

5.5 Conclusions

This investigation yields the following principal findings:

(1) This study employs DFN modeling to numerically analyze hydraulic fracture interactions at center- and edge-type intersections under variable crossed natural fractures and fluid injection rates. Through systematic variation of key sensitivity parameters (including the orientation, spacing, length, and persistence of pre-existing fractures, along with injection rate), some representative scenarios are established to quantify their influence on hydraulic fracture propagation dynamics.

(2) This study identifies center- and edge-type fracture network morphologies emerging under variable crossed natural fractures and fluid injection conditions. In edge-type propagation, hydraulic fractures intersecting natural fracture edges facilitate penetration into distal regions from perforation points; critically, when approach angles between hydraulic and natural fractures fall below critical thresholds, fracture reorientation activates natural fracture systems. Conversely, center-type propagation results from hydraulic fractures intersecting clustered natural fractures, forming radial fracture networks. Unlike large-scale features, this work reveals that small-scale aggregated intersections exhibit heightened sensitivity to both hydraulic fracture propagation behaviors (particularly in final morphology) and fluid injection rate variations.

(3) Fracture length exhibits a positive correlation with gas production during hydraulic fracturing, quantified through derived fitting curves. Center-type intersections of hydraulic fracture networks—governed by natural fracture sensitivity factors (orientation, spacing, length, persistence) and injection rates—generate extended fractures that enhance production potential. Consequently, hydrofracturing designs must strategically promote such intersections based on natural fracture morphology. When small-scale natural fractures form aggregated center-edge intersections, clustered low-pressure zones emerge without further production gains. Conversely, large-scale fractures produce striped low-pressure zones (Wang et al., 2021), precluding clustered zone formation. Thus, aggregated small-scale fractures may become counterproductive to production optimization under certain conditions.

This work examines planar fracture network interactions at center- and edge-type intersections under variable crossed natural fractures and fluid injection rates, offering fundamental insights for engineering applications. Future research will incorporate three-dimensional spatial characterization of natural fracture networks to investigate hydraulic fracture propagation and intersection behaviors in complex geological formations.

References

Cao, M., Hirose, S., & Sharma, M. M. (2022). Factors controlling the formation of complex fracture networks in naturally fractured geothermal reservoirs. *Journal of Petroleum Science and Engineering, 208*, 109642.

Cheng, W., Jin, Y., & Chen, M. (2015). Reactivation mechanism of natural fractures by hydraulic fracturing in naturally fractured shale reservoirs. *Journal of Natural Gas Science and Engineering, 23*, 431–439.

Dehghan, A. N., Goshtasbi, K., Ahangari, K., & Jin, Y. (2015). The effect of natural fracture dip and strike on hydraulic fracture propagation. *International Journal of Rock Mechanics & Mining Sciences, 75*, 210–215.

ELFEN TGR User and Theory Manual (2016). Rockfield Software Ltd, United Kingdom.

Fatahi, H., Hossain, M. M., & Sarmadivaleh, M. (2017). Numerical and experimental investigation of the interaction of natural and propagated hydraulic fracture. *Journal of Natural Gas Science and Engineering, 37*, 409–424.

Ghaderi, A., Taheri-Shakib, J., & Nik, M. A. S. (2018). The distinct element method (DEM) and the extended finite element method (XFEM) application for analysis of interaction between hydraulic and natural fractures. *Journal of Petroleum Science and Engineering, 171*, 422–430.

Guo, J., Zhao, X., Zhu, H., Zhang, X., & Pan, R. (2015). Numerical simulation of interaction of hydraulic fracture and natural fracture based on the cohesive zone finite element method. *Journal of Natural Gas Science and Engineering, 25*, 180–188.

Hou, B., Chen, M., Cheng, W., & Diao, C. (2016). Investigation of hydraulic fracture networks in shale gas reservoirs with random fractures. *Arabian Journal for Science and Engineering, 41*, 2681–2691.

Karimi-Fard, M., & Firoozabadi, A. (2001). Numerical simulation of water injection in 2D fractured media using discrete-fracture model. In *SPE annual technical conference and exhibition*, SPE-71615-MS.

Li, B. (2022). Modeling of shale gas transport in multi-scale complex fracture networks considering fracture hits. *Transport in Porous Media, 149*, 71–86.

Li, P., Dong, Y., Wang, S., & Li, P. (2020). Numerical modelling of interaction between hydraulic fractures and natural fractures by using the extended finite element method. *Advances in Civil Engineering, 2020*, 8848900.

Makedonska, N., Karra, S., Viswanathan, H., & Guthrie, G. (2020). Role of interaction between hydraulic and natural fractures on production. *Journal of Natural Gas Science and Engineering, 82*, 103451.

Rahman, M. M., & Rahman, S. S. (2013). Reactivation mechanism of natural fractures by hydraulic fracturing in naturally fractured shale reservoirs. *International Journal of Geomechanics, 13*, 809–826.

Sanchez, E. C. M., Cordero, J. A. R., & Roehl, D. (2020). Numerical simulation of three-dimensional fracture interaction. *Computers and Geotechnics, 122*, 103528.

Seifollahi, S., Dowd, P. A., Xu, C., & Fadakar, A. Y. (2014). A spatial clustering approach for stochastic fracture network modelling. *Rock Mechanics and Rock Engineering, 47*, 1225–1235.

Shakib, J. T., Akhgarian, E., & Ghaderi, A. (2015). The effect of hydraulic fracture characteristics on production rate in thermal EOR methods. *Fuel, 141*, 226–235.

Song, Y., Lu, W., He, C., & Bai, E. (2020). Numerical simulation of the influence of natural fractures on hydraulic fracture propagation. *Geofluids, 2020*, 8878548.

Suo, Y., Chen, Z., Rahman, S., & Yan, H. (2020). Numerical simulation of mixed-mode hydraulic fracture propagation and interaction with different types of natural fractures in shale gas reservoirs. *Environmental Earth Sciences, 79*, 279.

Taleghani, A., Gonzalez, M., & Shojaei, A. (2016). Overview of numerical models for interactions between hydraulic fractures and natural fractures: Challenges and limitations. *Computers and Geotechnics, 71*, 361–368.

Wang, Y., Li, X., & Tang, C. A. (2016). Effect of injection rate on hydraulic fracturing in naturally fractured shale formations: a numerical study. *Environmental Earth Sciences, 75*, 935.

Wang, S., Jiang, M., Dong, K., & Chen, M. (2018a). Fracture cross extension arithmetic research based on non-continuum theory. *Wireless Personal Communications, 103*, 41–53.

Wang, Y., Ju, Y., & Yang, Y. (2018b). Adaptive finite element-discrete element analysis for microseismic modelling of hydraulic fracture propagation of perforation in horizontal well considering pre-existing fractures. *Shock and Vibration, 2018*, 2748408.

Wang, Y., Ju, Y., Chen, J., & Song, J. (2019). Adaptive finite element-discrete element analysis for the multistage supercritical CO_2 fracturing of horizontal wells in tight reservoirs considering pre-existing fractures and thermal-hydro-mechanical coupling. *Journal of Natural Gas Science and Engineering, 61*, 251–269.

Wang, Y., Duan, Y., Liu, X., Huang, J., & Hao, N. (2021). Numerical analysis for dynamic propagation and intersection of hydraulic fractures and pre-existing natural fractures involving the sensitivity factors: Orientation, spacing, length, and persistence. *Energy & Fuels, 35*, 15728–15741.

Wasantha, P. L. P., Konietzky, H., & Weber, F. (2017). Geometric nature of hydraulic fracture propagation in naturally-fractured reservoirs. *Computers and Geotechnics, 83*, 209–220.

Xiong, D., & Ma, X. (2022). Influence of natural fractures on hydraulic fracture propagation behavior. *Engineering Fracture Mechanics, 276*, 108932.

Yan, X., & Yu, H. (2022). Numerical simulation of hydraulic fracturing with consideration of the pore pressure distribution based on the unified pipe-interface element model. *Engineering Fracture Mechanics, 275*, 108836.

Zhao, H., Li, W., Wang, L., Fu, J., Xue, Y., Zhu, J., & Li, S. (2022). The influence of the distribution characteristics of complex natural fracture on the hydraulic fracture propagation morphology. *Frontiers in Earth Science, 9*, 784931.

Zhou, J., Zhang, L., Braun, A., & Han, Z. (2017). Investigation of processes of interaction between hydraulic and natural fractures by PFC modeling comparing against laboratory experiments and analytical models. *Energies, 10*(7), 1001.

Chapter 6
Wells Connection and Long Hydraulic Fracture Induced by Multi-Well Hydrofracturing Utilizing Crossed Perforation Clusters

6.1 Introduction

The hydraulic fracturing of horizontal wells is an important technique for exploiting shale gas and tight gas in deep unconventional reservoirs. A high-pressure fracturing fluid is injected into the perforation mainly through a segmented perforation process, and the rock mass is fractured to form the main fracture and induce the formation of multi-branch fractures. The fracturing fractures are connected to the natural fractures to form a complex fracture network, through which the oil and gas reservoirs can direct the flow into horizontal wells in order to achieve fracturing and stimulation. Complexity of the fracture network is the key to increasing production (Wang et al., 2020).

Multi-well hydrofracturing is affected by the stress shadows of multiple fractures in other horizontal wells as well as in the same horizontal well, resulting in a more complex fracture network than found in a single well (Liu et al., 2020). Fracture interference between wells caused by pressure sinks as well as the spatial and temporal evolution of in-situ stress magnitude and azimuth caused by pressure depletion becomes an additional challenge for reservoir development (Marongiu-Porcu et al., 2016). During horizontal well fracturing, the formation between fractured well and adjacent horizontal well is directly connected with fracturing fractures. This phenomenon is called horizontal well-fracturing channelling (Guo et al., 2016a). Well-fracturing channelling and fracture hits are well-to-well communication events caused by fracturing that can result in production losses (or gains) and mechanical damage (Guo et al., 2016b; King et al., 2017). Owing to the disturbance caused by the connection of fracturing fractures, hydraulic fractures form a long single-wing fracture feature under the influence of geological characteristics. This leads to a longer propagation of fracturing fractures, which are connected with the fracturing fractures of adjacent wells, leading to fracture hits (Esquivel & Blasingame, 2017). Fracture

Y. Wang, *Deflection of Fracturing Fracture Network Disturbed by Discontinuity and Multiple Fractures in Rock*, https://doi.org/10.1007/978-981-95-1569-1_6

hits are particularly common in closely horizontal wells and usually occur in fracture planes or complex locations (King et al., 2017). These connections have also been described as direct hydraulic connections or stress shadow effects, often resulting in fracture growth imbalance (Daneshy & Pomeroy, 2012; Daneshy et al., 2015). One of the most direct factors affecting fracture hits in horizontal well fracturing is the well pattern and spacing (Lin et al., 2013). Guo et al. (2016a) used numerical simulations of low-permeability reservoirs in the presence of high-conductivity fractures to compare the impact of different well spacings on productivity and optimised reasonable well spacing to ensure maximum productivity and prevent fracturing channelling (Guo et al., 2016a). It is also necessary to further study the qualitative and quantitative relationships between fracturing channelling and geological factors, such as in-situ stress, formation, and the fracture (Guo et al., 2016b). Guo et al. (2016a), (2016b) found that increasing the number of fracturing stages and well spacing in cross-horizontal wells could not only effectively reduce the fracturing channelling rate but also ensure productivity requirements. Numerical simulation considering the stress field and formation pressure variation is one of the main methods for evaluating pressure channelling between wells (Guo et al., 2016a).

The perforation cluster spacing, initiation sequence, and well spacing in a well factory fracturing process composed of multiple horizontal wells are the main factors that cause different degrees of unstable propagation of parallel fractures (Izadi et al., 2015; Ju et al., 2020). Currently, multi-well hydrofracturing sequences mainly include sequential, simultaneous, alternate, and zipper fracturing (crossed perforation clusters) (Jo, 2012; Liu et al., 2020; Qiu et al., 2015; Yu & Sepehrnoori, 2013). Simultaneous fracturing refers to the simultaneous injection of fluids into all the perforation clusters of several horizontal wells (Liu et al., 2020). Fracture interactions between wells limit the fracture width and shorten the fracture length. In a study by Liu et al. (2020), the total fracture length of zipper fracturing (crossed perforation clusters) was longer than those of simultaneous and sequential fracturing, which is the best method for volume fracturing. Zipper fracturing (crossed perforation clusters) exhibits the best fracturing effect, followed by sequential fracturing (Jo, 2012). The change in the well spacing affects the unstable propagation of the fractures. For two horizontal wells under different well spacings, the main fractures formed by the fractures of the two horizontal wells along the principal stress direction will hinder the development of other fractures, and as the well spacing increases, the fracture width at the completion of fracturing increases (Duan et al., 2021; He et al., 2020). When the well spacing is too small, serious interference occurs between multiple wells, and also the fracture tips between multiple wells are connected, affecting the complexity of the fracture network. If the well spacing is too large, natural fractures in some areas between multiple wells cannot be activated, thereby affecting oil and gas production (Duan et al., 2021; He et al., 2020; Li & Zhang, 2018). The total length of hydraulic fractures decreases with a decrease in well spacing, and the total volume of the hydraulic fractures increases with a decrease in well spacing (Wang &

Liu, 2022). In addition to the well spacing, the fracture propagation and interaction under different initiation sequences of multiple wells are also different.

In this study, the generation of long hydraulic fractures and well fracture hits in multi-well hydrofracturing were analysed by comparing the hydraulic fracture propagation and dynamic evolution of the stress field during fracture propagation under multi-well cross and parallel perforation clusters.

6.2 Governing Equations of Multi-Well Hydrofracturing Considering Thermal–Hydraulic-Mechanical Coupling

The stratal deformation, fluid seepage, fracture fluid flow, and heat transfer are coupled and considered in the model in this study to form thermal-hydro-mechanical coupling. The geomechanical equations of stratal movement and microseismic analysis induced by multi-well hydrofracturing are provided as below (Wang, 2020; Wang et al., 2019).

$$\text{Stratal deformation}: \quad \boldsymbol{L}^T(\boldsymbol{\sigma}^e - \alpha \boldsymbol{m} p_s) + \rho_b \boldsymbol{g} = 0 \tag{6.1}$$

$$\text{Fluid seepage}: \quad \text{div}\left[\frac{k}{\mu_l}(\nabla p_l - \rho_l g)\right] = \left(\frac{\phi}{K_l} + \frac{\alpha - \phi}{K_s}\right)\frac{\partial p_l}{\partial t} + \alpha \frac{\partial \varepsilon_v}{\partial t} \tag{6.2}$$

$$\text{Fracture fluid flow}: \quad \frac{\partial}{\partial \boldsymbol{x}}\left[\frac{k^{fr}}{\mu_n}(\nabla p_n - \rho_{fn}\boldsymbol{g})\right] = S^{fr}\frac{\mathrm{d}p_n}{\mathrm{d}t} + \alpha(\Delta \dot{e}_\varepsilon) \tag{6.3}$$

$$\text{Heat transfer}: \quad \text{div}\left[k_b \nabla T_f\right] = \rho_b c_b \frac{\partial T_f}{\partial t} + \rho_f c_f q_f \nabla T_f \tag{6.4}$$

The involved symbols and physical meaning of parameters are summarised in Table 6.1. In this study, the combined finite element-discrete element method was used to solve the governing equations and analyse the fracture propagation process (Wang et al., 2019).

Table 6.1 Symbols and physical meaning of parameters

Symbol	Physical meaning	Symbol	Physical meaning
$\boldsymbol{L}$	Spatial differential operator	k_b	Thermal conductivity
$\boldsymbol{\sigma}^e$	Effective stress tensor	T_f	Fluid temperature
$\boldsymbol{m}$	Identity tensor	$\boldsymbol{q}_f$	Darcy fluid flux
$\boldsymbol{g}$	Gravity vector	t	Time
ε_v	Volumetric strain of the rock formation	$\Delta \dot{e}_E$	Aperture strain rate

6.3 Numerical Models of Multi-Well Hydrofracturing Utilizing Crossed Perforation Clusters

In this study, an engineering-scale initial geometric model of multi-well multi- perforation staged fracturing in a deep tight rock mass was established, as shown in Fig. 6.1. Three horizontal wells (denoted as Wells 1, 2, and 3) were set in this model, and five perforation clusters (numbered 1–5 in sequence) were set for each well. There are two geometric variables in the model: a is the perforation cluster spacing and b is the well spacing. In Fig. 6.1a, multiple well perforation clusters are set in parallel, and in Fig. 6.1b, multiple well perforation clusters are crossed. Figure 6.2 shows a mesh subdivision of the model with an initial dense mesh for the local area of each perforation to obtain a reliable fracture propagation path. The case settings of different perforation cluster distributions and well spacings (b = 100, 75, 50, and 25 m) are listed in Table 6.2. The perforation clusters in each well for the cases listed in Table 6.2 were subjected to sequential fracturing. Table 6.3 shows the case settings for different perforation distributions and well initiation sequences. Interpretations of different fracturing sequences can be found in Wang and Liu (2022). In all cases, the fracturing fluid temperature of all cases are set at 20 °C, rock matrix temperature is set at 60 °C, and the perforation cluster spacing are set at 75 m. The basic physical parameters of the model were set as shown in Table 6.4 and were tested on tight rock samples in the Shengli Oilfield in Shandong Province, China.

6.4 Results and Analysis

6.4.1 Hydraulic Fracture Propagation of Parallel and Crossed Perforation Clusters in Multi-Wells

Cases I and IV were considered as examples to analyse the fracture propagation and stress field evolution of different perforation cluster arrangements. Figures 6.3 and 6.4 show the final fracture distribution morphology and displacement (unit: m) in the x direction of the multi-well parallel and crossed perforation clusters. In Fig. 6.3, owing to the effect of fracturing of Wells 1 and 2 on the reservoir in-situ environment, all fractures in Well 3 are deflected, and the long fracture 3 in Well 2 is connected to a part of the perforation in Well 3, resulting in fracture hits between wells and a short propagation length of fracture 3. In contrast to Fig. 6.3, under the multi-well crossed perforation clusters, the effect of Well 1 fracturing on the in-situ reservoir does not lead to the connection of fractures between Wells 2 and 1. Also, because of the crossed perforation clusters, longer fractures do not connect with other perforation clusters, and well fractures are greatly reduced. Each fracture propagates independently.

Figures 6.5 and 6.6 show the evolution results of the fracture network propagation and shear stress (unit: MPa) at t = 2502 s, t = 5002 s and t = 7502 s for parallel

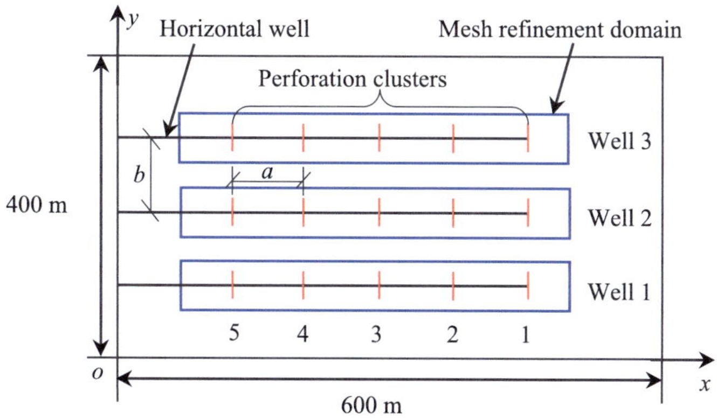

(a) Multi-well parallel perforation clusters

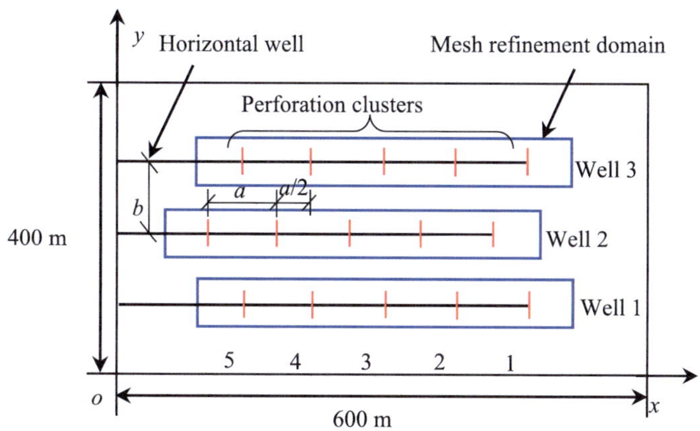

(b) MuHi-well cross-perforation clusters

Fig. 6.1 Initial geometric model of multiple wells multistage fracturing

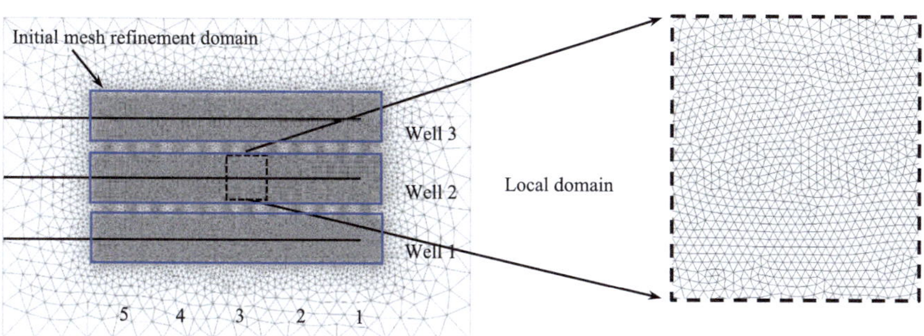

Fig. 6.2 Initial mesh refinement of finite element model

Table 6.2 Case setting of different perforation cluster distribution and well spacings

Cases	Multi-well perforation cluster distribution	Well initiation sequences	Well spacing b/m	Fracturing fluid temperature/°C	Rock matrix temperature/°C
I	Parallel	Sequential	75	20	60
II	Cross	Sequential	25	20	60
III	Cross	Sequential	50	20	60
IV	Cross	Sequential	75	20	60
V	Cross	Sequential	100	20	60

Table 6.3 Case setting of different well initiation sequences

Cases	Multi-well perforation cluster settings	Well initiation sequences	Well spacing b/m	Fracturing fluid temperature/°C	Rock matrix temperature/°C
VI	Cross	Sequential	75	20	60
VII	Cross	Simultaneous	75	20	60
VIII	Cross	Alternate 1 → 3 → 2	75	20	60
IX	Cross	Alternate 2 → 1 → 3	75	20	60

Table 6.4 Basic physical parameters of the numerical model

Parameters	Value
Horizontal minimum in-situ stress (x-direction) S_h/MPa	40
Horizontal maximum in-situ stress (y-direction) S_H/MPa	44
Fluid injection rate Q/(m^3/s)	0.5
Pore pressure p_s/MPa	10
Biot coefficient α	0.75
Elastic modulus E/GPa	31
Poisson ratio ν	0.22
Penetration k/(nD)	50
Porosity ϕ	0.05
Kinematic viscosity coefficient μ_n/(Pa · s)	1.67×10^{-3}
Fracture fluid bulk modulus K_f^{fr}/MPa	2000
Tensile strength σ_t/MPa	5.26
Fracture energy G_f/(N · m)	165

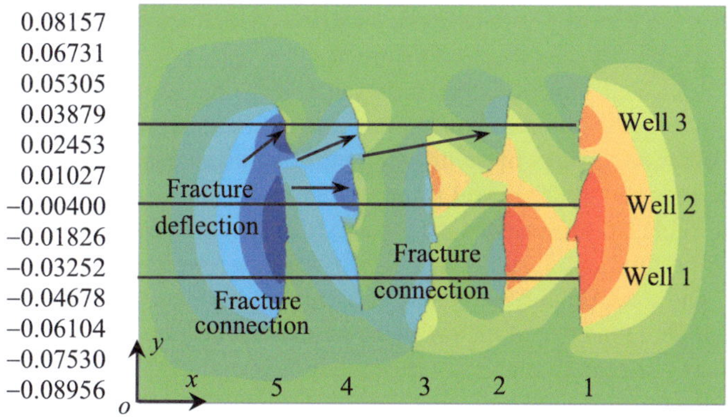

Fig. 6.3 Final fracture morphology and displacement of multi-well parallel perforation clusters

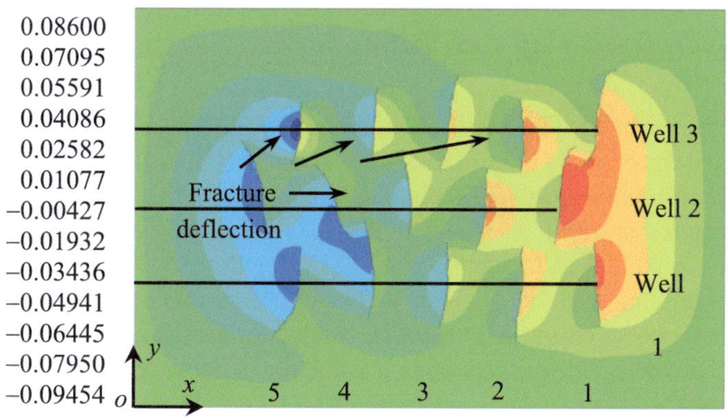

Fig. 6.4 Final fracture morphology and displacement of multi-well crossed perforation clusters

and crossed perforation clusters in multiple wells, respectively. Under the crossed perforation clusters, the stress shadow area between the two adjacent wells decreased, and the fracture deflection and connection caused by the shear stress disturbance were evidently weakened. As shown in Fig. 6.6b, after fracturing Well 1, superposition of the shear stress occurred in the local area adjacent to Wells 1 and 2, resulting in a stress disturbance that deflected the fracture. Simultaneously, the stress zone of Well 1 disturbed the formation of a stress shadow zone between multiple wells. When the shear stresses of fractures 2 and 3 were superimposed, the shear stresses on both sides of the fracture tip were symmetrical without any obvious deflection. As seen in Fig. 6.6c, in the local areas of Wells 2 and 3, the shear stresses are superimposed and reduced owing to the influence of the stress shadow, and a larger deflection occurred in fracture 1 of Well 3. After the perforation cross-arrangement of the three wells, the stress shadow area around the fracture was reduced, the disturbance of the stress field was weakened, and the shear stresses on both sides of the upper and lower tips of the fracture were mostly symmetrically distributed. A fracture propagation pattern with single fracture development was formed.

Taking multi-well sequential fracturing at a 75 m well spacing as an example, and by comparing the different perforation cluster distributions, the arrangement of crossed perforation clusters in multi-well hydrofracturing can alleviate the fracture interference between multiple wells and ensure that a single fracture can be better propagated. Table 6.5 summarizes the total fracture length and volume of the parallel and crossed perforation clusters under 75 m well spacing in multi-well sequential fracturing. The total fracture length of the crossed perforation clusters was more than 100 m higher than that of the parallel perforation clusters. However, the volume of the crossed perforation clusters was smaller than that of the parallel perforation clusters.

6.4.2 Connected Long Hydraulic Fractures in Multi-Well Hydrofracturing with Different Well Spacings

Figure 6.7 shows the final fracture network morphology and shear stress distribution for sequential fracturing with different well spacings (Cases II, III, IV and V). The influence of different well spacings on long fractures in multi-well fracture propagation was analysed in detail. As shown in Fig. 6.7a, when the well spacing was 25 m, the fractures in Wells 1 and 3 were connected during the propagation process owing to the small spacing, and the fractures in Well 2 propagated to the other two wells. The fracture connection between multiple wells produces long fractures, which is not conducive to the generation of complex fracture network, and to the exploitation of oil and gas in practical engineering. Small well spacing also results in excessive deflection of the fractures. Compared with Wells 1 and 2, the fracture propagation in Well 3 was significantly inhibited. The lower end of the fracture stopped after contact with the fracture in Well 1, and the upper end of the fracture stopped prematurely. At 50 m spacing, the fracture spacing between adjacent wells increases, the deflection phenomenon of the fracture tip is significantly alleviated, and there is no connection between the fractures, which reduces the occurrence of long fractures. The fractures in Well 3 were severely inhibited, and the deflection of fractures 1–4 in Well 3 increased. Owing to the long fracture 5 in Well 1 (the tip of fracture 5 propagated to the area between Wells 2 and 3), the deflection of fracture 5 in Well 2 intensified, and the length of fracture 5 in Well 3 was inhibited. When the well spacing increased to 75 m, the interference between the fractures in multiple wells was reduced, and the deflection of the fractures in Well 3 weakened; however, fracture 5 was shorter and severely inhibited. When the well spacing was 100 m, most fractures propagated along the initial perforation direction (y-direction). Only fractures 1 and 2 in Wells 1 and 2 and fractures 1 and 2 in Well 2 were deflected at the fracture tip, and the deflection angle was not large.

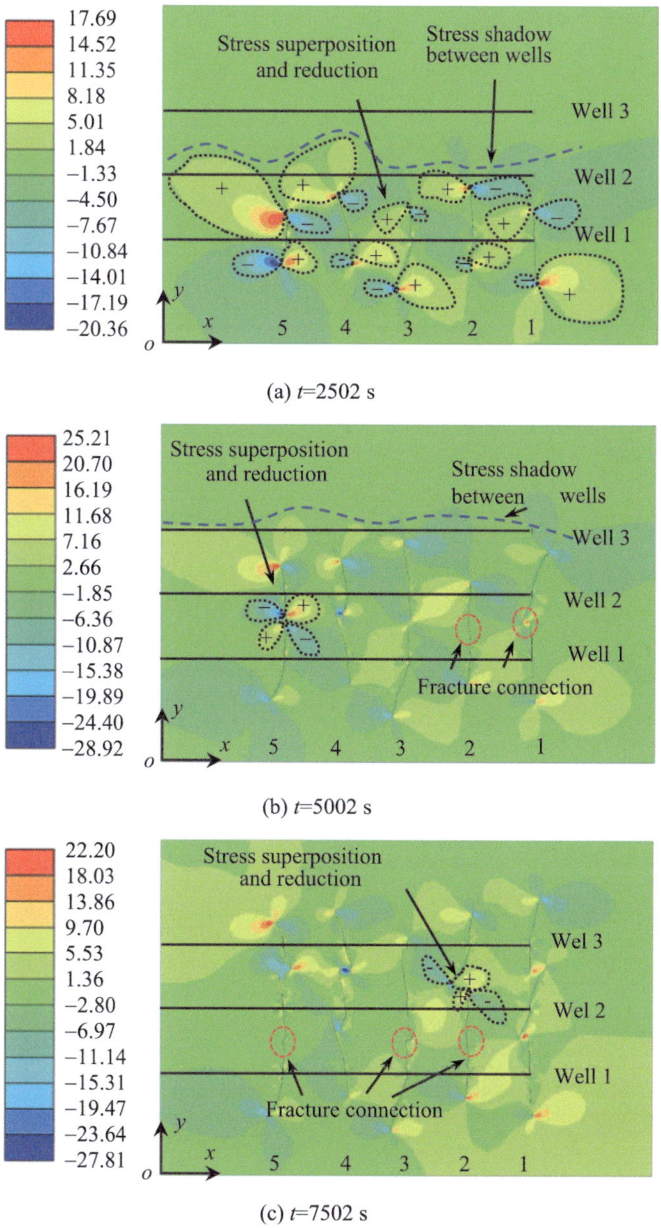

Fig. 6.5 Evolution of shear stress τ_{xy} (MPa) in multi-well hydrofracturing of parallel perforation clusters

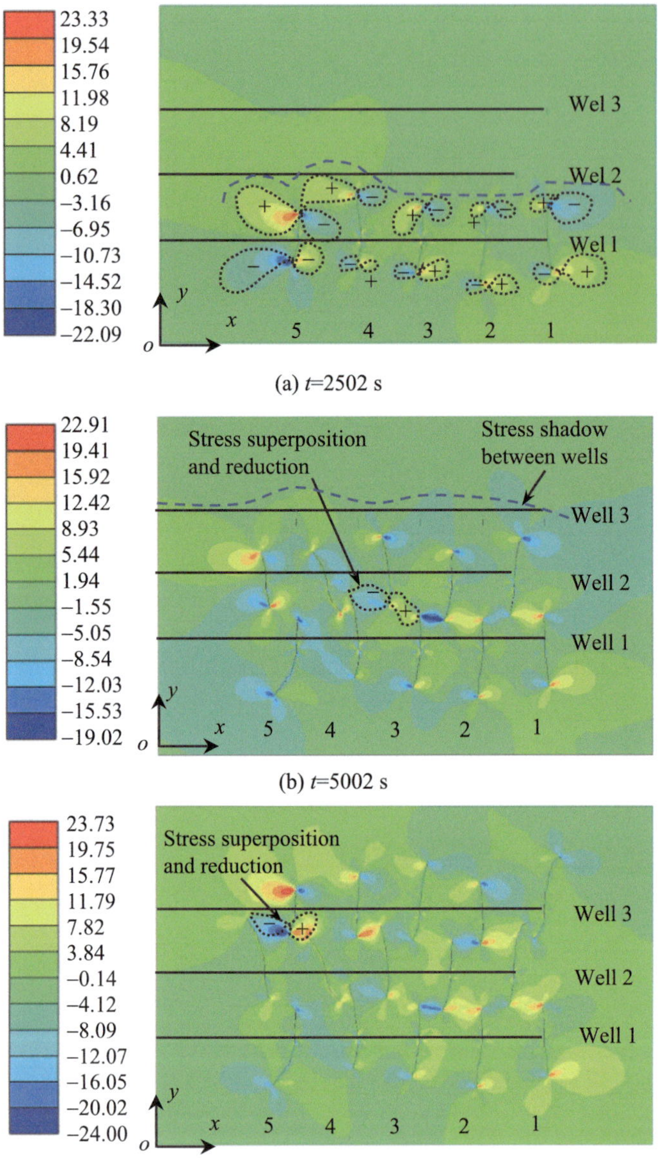

(a) t=2502 s

(b) t=5002 s

(c) t=7502 s

Fig. 6.6 Evolution of shear stress τ_{xy} (MPa) in multi-well hydrofracturing of crossed perforation clusters

Table 6.5 Total fracture length and volume of parallel and crossed perforation clusters in multi-well hydrofracturing

Case	Perforation cluster distribution	Total fracture length *L*/m	Total fracture volume V/m^3
I	Parallel	1271.25	595.70
IV	Cross	1390.65	581.21

In the sequential fracturing process of the three horizontal wells, the fracture produced a certain deflection, and the fracture interference between the wells was the most serious when the well spacing was 25 m. When the well spacing is small, the fractures are connected, and fracture propagation is inhibited. With an increase in well spacing, the fracture interference became weaker, the connected fractures were reduced, and the degree of fracture deflection decreased. Table 6.5 summarizes the total fracture lengths and volumes of different well spacings in multi-well hydrofracturing. To analyze the relationship between the fracture length, volume, and well spacing more intuitively, Table 6.6 is drawn as a line graph, as shown in Fig. 6.8. With an increase in well spacing, the total fracture length gradually increased, and the total fracture volume gradually decreased. However, as the well spacing reached 75 m, an increase in well spacing resulted in a decrease in the fracture volume. Therefore, to achieve a large fracture length and volume simultaneously, it is necessary to select an appropriate well spacing that is more conducive to obtaining a better fracturing effect and improving reservoir production.

6.4.3 Connected Long Hydraulic Fractures in Multi-Well Hydrofracturing with Different Well Initiation Sequences

Figure 6.9 shows the final fracture network morphology and shear stress distribution in different well initiation sequences (Case VI, VII, VIII and IX) at 75 m well spacing. Under sequential fracturing, the lower half of fracture 5 in Well 1 deflected slightly, and the tips of fractures 1, 4, and 5 in Well 2, whereas the tip of fracture 1 in Well 3 deflected to different degrees along the x-axis. Among them, the tip of fracture 5 in wells 2 and 3 deflected to nearly 90° owing to interference. Under simultaneous fracturing, during the process of sequential initiation and propagation of perforation in multiple wells, the subsequent initiation fractures interfere with and are deflected by the right fractures. The fractures in Wells 1 and 3 were severely deflected from Well 2 at the same time, the lengths of fractures 2, 3, 4, and 5 in Well 2 and fracture 5 in Wells 1 and 3 were severely inhibited owing to the interference of the fractures on both sides. In alternate fracturing (Case VIII), as the Wells 1 and 3 are first fractured, fractures 2 and 3 in Wells 1 and 3 propagate longer, which is caused by the large spacing between Well 1 and Well 3 (equivalent to twice the well spacing). Therefore,

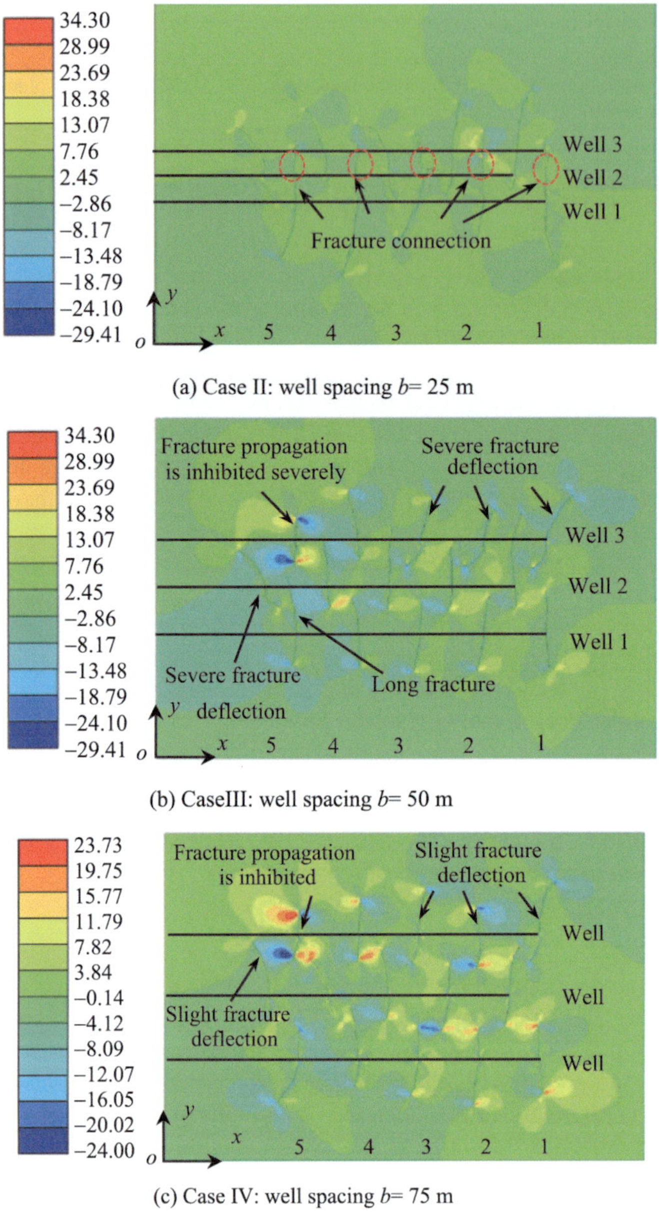

Fig. 6.7 Final fracture morphology and shear stress τ_{xy} (MPa) under different well spacings in sequential fracturing of crossed perforation clusters

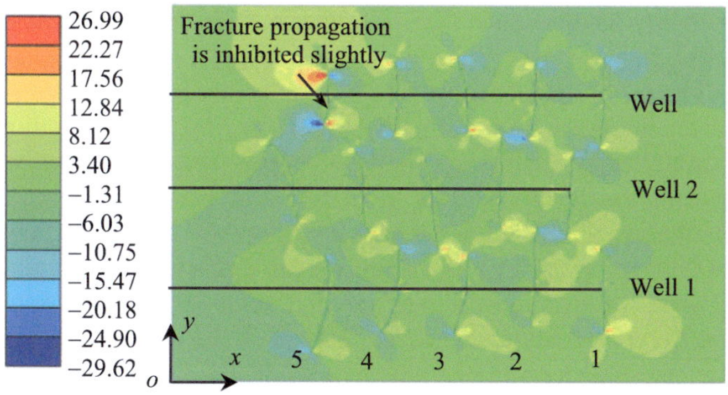

(d) Case V: well spacing b= 100 m

Fig. 6.7 (continued)

Table 6.6 Total fracture length and volume of different well spacings in multi-well hydrofracturing

	Well spacing b/m	Total fracture length L/m	Total fracture volume V/m^3
Case II	100	1401.20	577.11
Case III	75	1390.65	581.21
Case IV	50	1373.35	570.79
Case V	25	1358.88	575.75

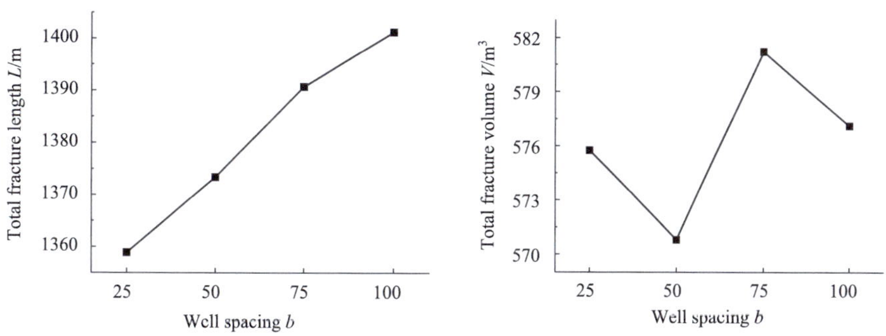

(a) Total fracture length (b) Total fracture volume

Fig. 6.8 Total fracture length and volume under different well spacings of crossed perforation clusters

the interference between Well 1 and Well 3 is weakened, and the fracture tips have slight deflection. Fractures in Well 2 were inhibited by fractures in Wells 1 and 3, and the fractures were short having almost no deflection, propagating completely along the initial perforation direction. In the alternate fracturing (Case IX), Well 2 was fractured first, and the fracture length in Well 2 increased compared with other well initiation sequences. Conversely, the corresponding fracture propagation in Wells 1 and 3 was slightly inhibited, and the tips of the fractures in Wells 1 and 3 exhibited a slight deflection.

The well initiation sequence had a significant influence on the fracture morphology and fracture length. In the case of simultaneous fracturing, the propagation length of Well 2 was inhibited by the fractures in the other two wells. Additionally, fracture 5 of the three wells was disturbed by the previous fractures, and the length was severely inhibited. In the two alternate fracturing cases, only the fracture length of Well 2 in sequential fracturing (Case IV) was slightly inhibited, and the remaining fractures propagated well. To quantitatively analyze and compare the fracture propagation lengths of different initiation sequences, Table 6.7 summarizes the total fracture propagation lengths under different initiation sequences. Compared with sequential fracturing, the fracture propagation length of alternate fracturing (Case VIII and IX) improved, and the improvement of alternate fracturing 2 → 1 → 3 (Case IX) was the most obvious. However, the total fracture propagation length of simultaneous fracturing was the smallest, which was not conducive to the increase in production. Figure 6.10 shows the total fracture length and volume results for different initiation sequences. The fracture length under simultaneous fracturing is the minimum, and the fracture length under alternate fracturing 2 → 1 → 3 (Case IX is the maximum. The fracture volume under alternate fracturing 1 → 3 → 2 (Case VIII) is the minimum, while the fracture volume under simultaneous fracturing is the maximum. This is because the fracture length was shorter, and it was easier for the fracturing fluid to accumulate in the fracture and form a larger fracture volume.

6.4.4 Multi-Well Hydrofracturing Induced Microseismic Events

6.4.4.1 Different Well Spacings

Microseismic events caused by hydraulic fracturing in multiple wells, including contact slip and damage events, are closely related to the interactions between fractures and the stress shadow effect. Figure 6.11 shows the distribution of damage and contact slip events for a single fracture at the same well spacing. Under the same well spacing, the number of damage events was much larger than the number of contact slip events, and most of the damage events occurred at the location of fracture propagation and were distributed along the location of the fracture propagation. Almost no damage occurred at the initial perforation, and most contact slip events occurred at the initial perforation position and at some locations with serious deflection. Figure 6.12 shows the distribution of damage and contact slip events under different well spacings. By comparing the number of damage and contact slip events under different well spacings, it can be found that the number of damage and contact

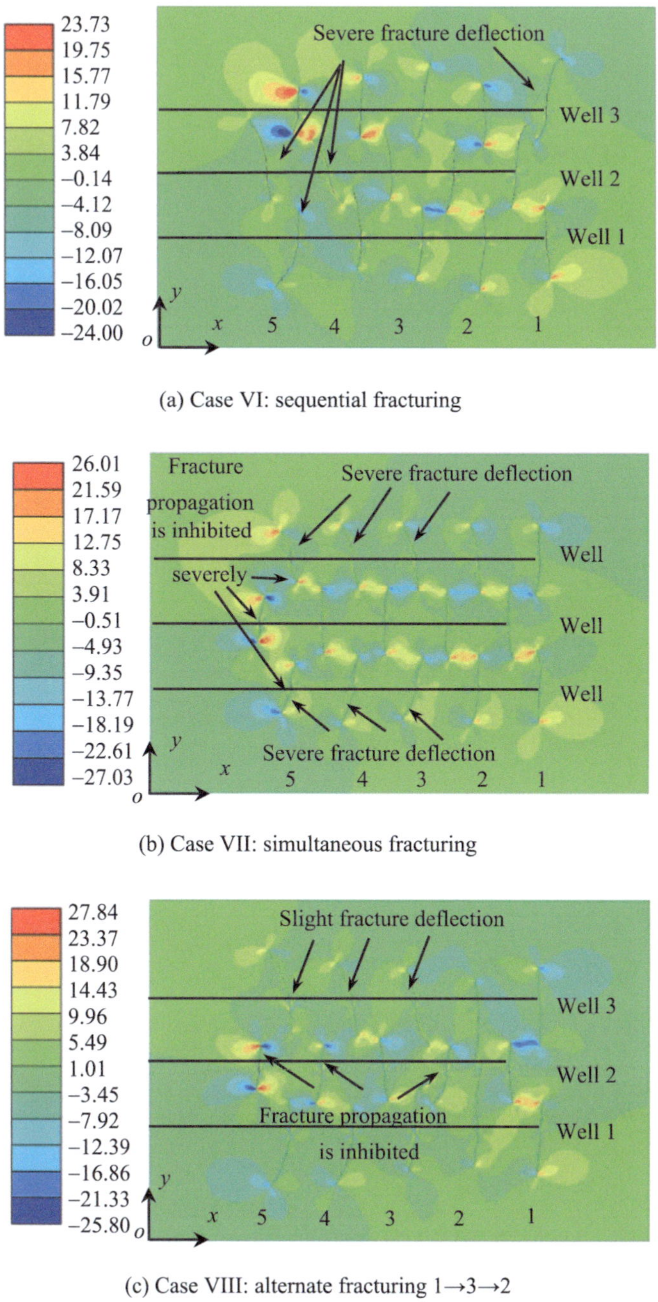

Fig. 6.9 Final fracture morphology and shear stress τ_{xy} (MPa) under different initiation sequences of crossed perforation clusters

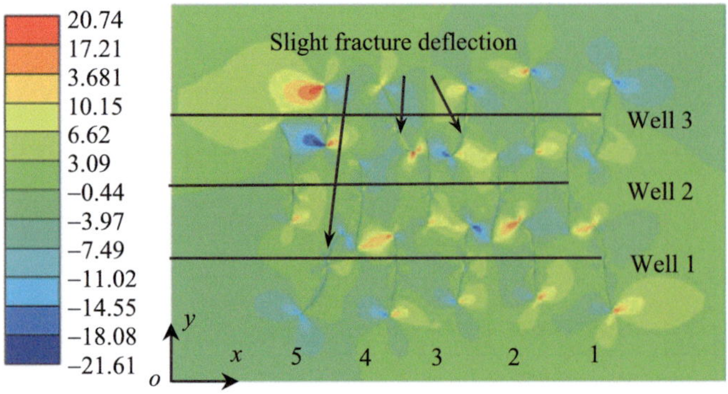

(d) Case IX: alternate fracturing 2→1→3

Fig. 6.9 (continued)

Table 6.7 Total fracture length and volume of different initiation sequences in multi-well hydrofracturing

	Initiation sequence	Total fracture length L/m	Total fracture volume V/m^3
Case VI	Sequential fracturing	1390.65	581.21
Case VII	Simultaneous fracturing	1214.52	600.19
Case VIII	Alternate fracturing 1 → 3 → 2	1394.03	575.86
Case IX	Alternate fracturing 2 → 1 → 3	1401.90	576.98

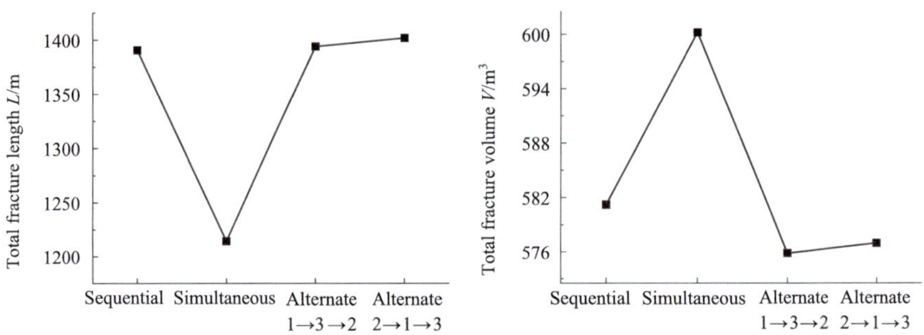

(a) Total fracture length.

(b) Total fracture volume.

Fig. 6.10 Total fracture length and volume under different initiation sequences of cross- perforation clusters

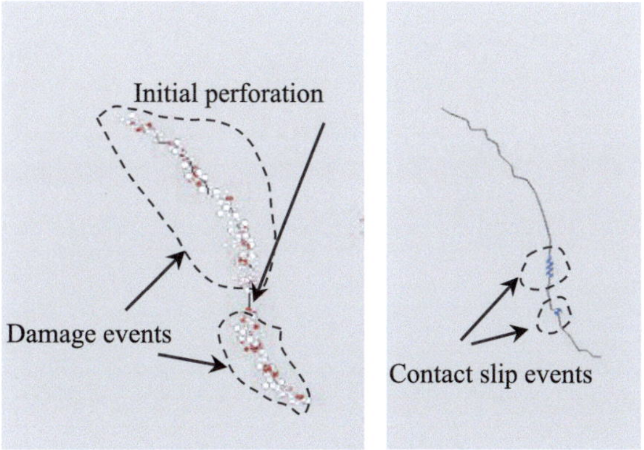

Fig. 6.11 Damage events and contact slip events for a single fracture

slip events increases with a decrease in the horizontal well spacing, which is due to the acceleration of the stress field change when the well spacing decreases. In sequential fracturing, owing to the influence of stress accumulation, fractures in the later stages of fracturing may cause more contact slip events, which is consistent with the analysis results of the evolution and disturbance of the shear stress field. As the well spacing decreased, the interaction between fractures and the shadow effect of shear stress became more serious, and the microseismic events formed around the fractures also increased.

Figure 6.13 shows the maximum and cumulative magnitude changes with time at different well spacing values. When the well spacing was 25 m, the maximum and cumulative magnitudes fluctuated between 0.0 and −4.5. When the well spacing is 50 m, the maximum magnitude and cumulative magnitude fluctuate between 0 and −5. Because of the small well spacing, fracturing increases the formation movement and stress changes between the fractures, thereby inducing higher levels of damage and contact slip events. The maximum magnitude and cumulative magnitude of the 75 and 100 m well spacing are similar because larger well spacing may not cause strong formation movement and stress changes. Therefore, the number and size of microseismic events are proportional to the formation movement and compression effects.

6.4.4.2 Different Initiation Sequences of Perforation Clusters

Figure 6.14 shows the distribution of the damage and contact slip events under different initiation sequences of the perforation clusters. The damage events were all around the location of the fracture propagation, and there was no obvious difference owing to the large number of dense distributions. The number of contact slip events generated by simultaneous fracturing was the lowest, which may have been due to the stress change in the fracture accumulation area. The contact slip events produced by alternate fracturing of 1 → 3 → 2 and 2 → 1 → 3 (Cases VIII and IX) were more

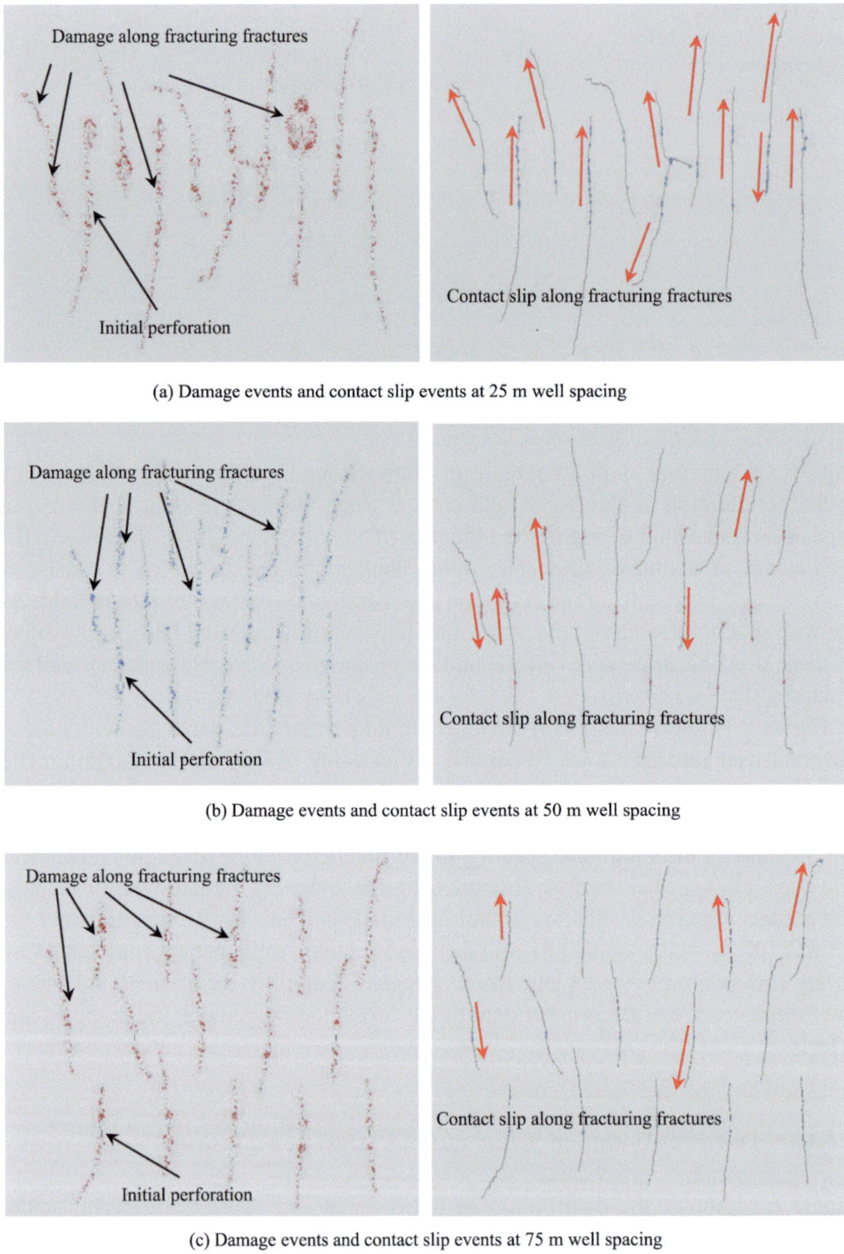

Fig. 6.12 Damage events and contact slip events distribution at different well spacings

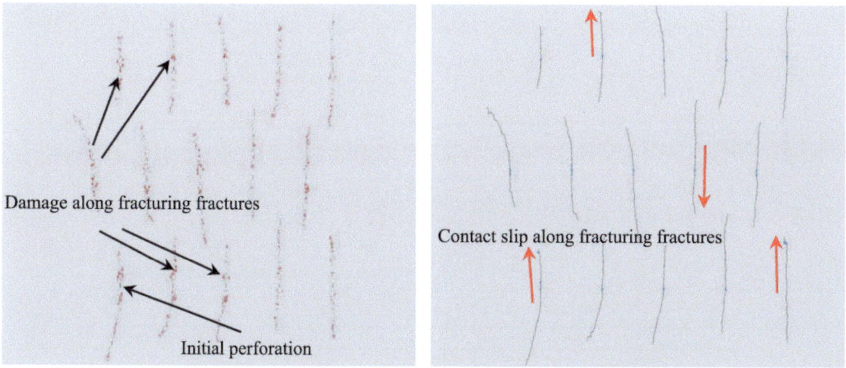

(d) Damage events and contact slip events at 100 m well spacing

Fig. 6.12 (continued)

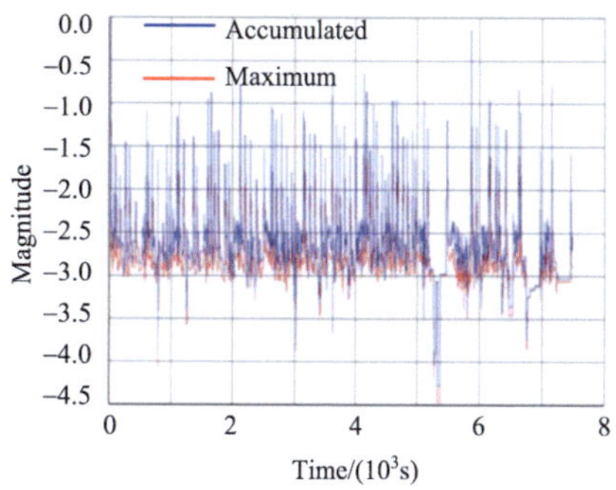

(a) Well spacing b= 25 m

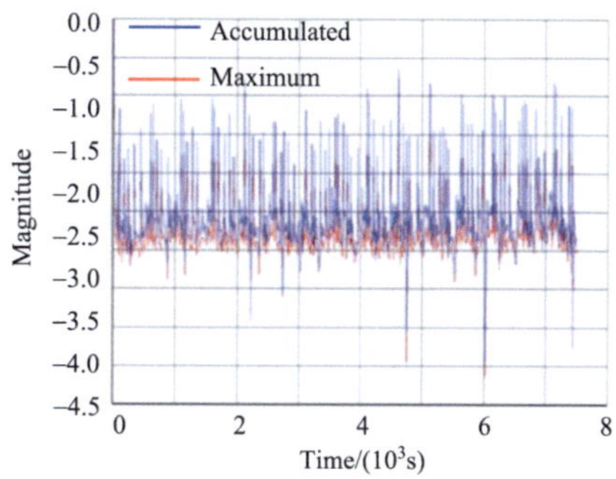

(b) Well spacing b= 50 m

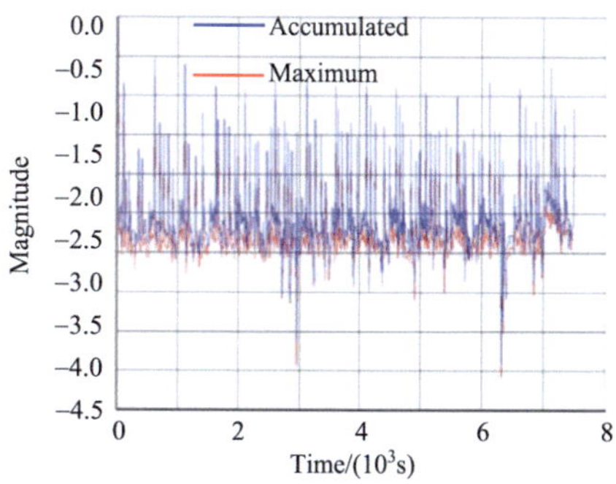

(c) Well spacing b= 75 m

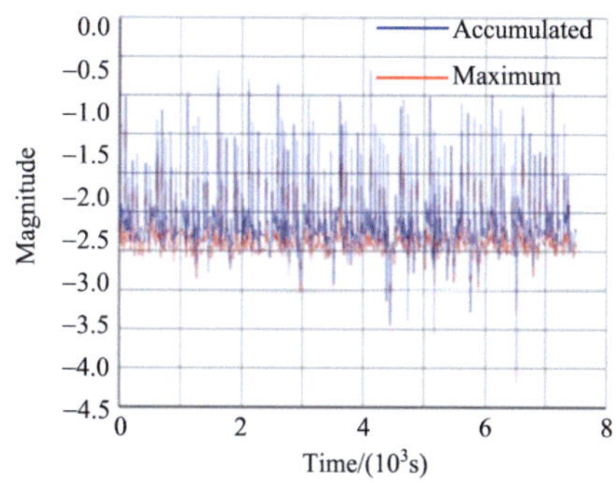

(d) Well spacing b= 100 m

Fig. 6.13 Maximum magnitude and cumulative magnitude change with time at different well spacings

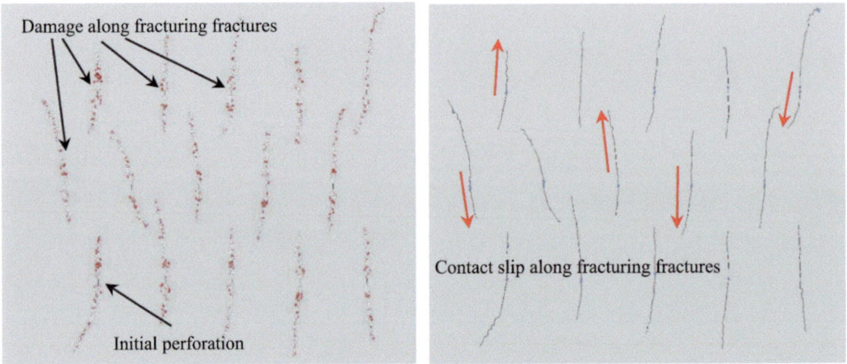

(a) Case VI: sequential fracturing

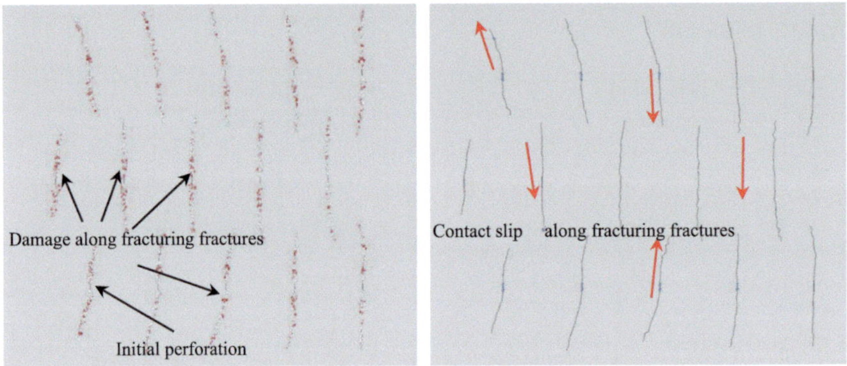

(b) Case VII: simultaneous fracturing

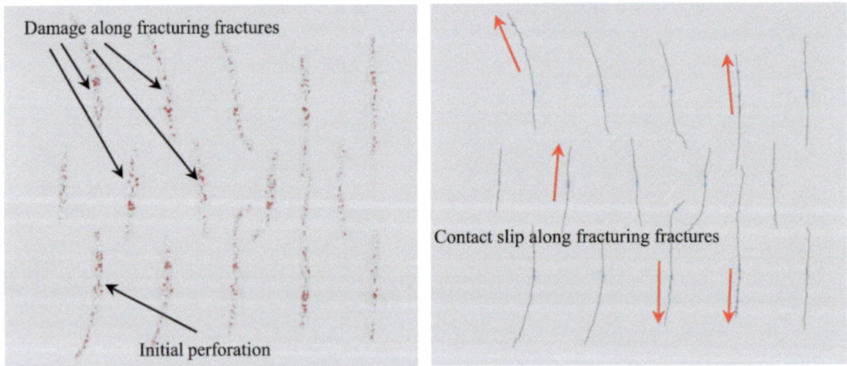

(c) Case VIII: alternate fracturing 1→3→2

Fig. 6.14 Damage events and contact slip events distribution under different initiation sequences

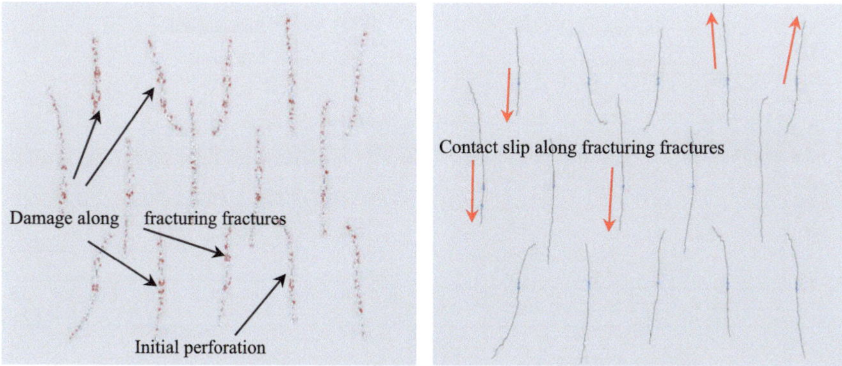

(d) Case IX: alternate fracturing 2→1→3

Fig. 6.14 (continued)

than those produced by simultaneous fracturing and less than those produced by sequential fracturing. For sequential fracturing, owing to the cumulative formation movement and compression, it was easier to produce more contact slip events than for simultaneous and alternate fracturing.

Figure 6.15 shows the maximum and cumulative magnitude changes with time for different initiation sequences. As shown in the figure, the maximum magnitude and cumulative magnitude in sequential fracturing fluctuate between 0 and −5, whereas in simultaneous fracturing fluctuate between 0 and −3.6, in alternate fracturing 1 → 3 → 2 (Cases VIII) fluctuate between 0 and −4, and in alternating fracturing 2 → 1 → 3 (Case VIII) fluctuate between 0 and −4.5. The fluctuation amplitude of simultaneous fracturing is the smallest, and the fluctuation amplitude of alternate fracturing 1 → 3 → 2 and 2 → 1 → 3 (Cases VIII and IX) is also lower than that of sequential fracturing. Fewer contact slip events occurred, and their amplitudes were relatively low. In addition, the magnitude of microseismic events caused by multistage fracturing is closely related to the related formation movement and compression. Large and diverse stratigraphic movements and compression effects markedly increase the likelihood of microseismic events and easily induce large-scale microseismicity (Wang et al., 2018).

6.5 Conclusions

In this study, the behaviours of long hydraulic fracture propagation and well fracture hits under parallel and crossed perforation clusters were compared and analysed. The conclusions are as follows.

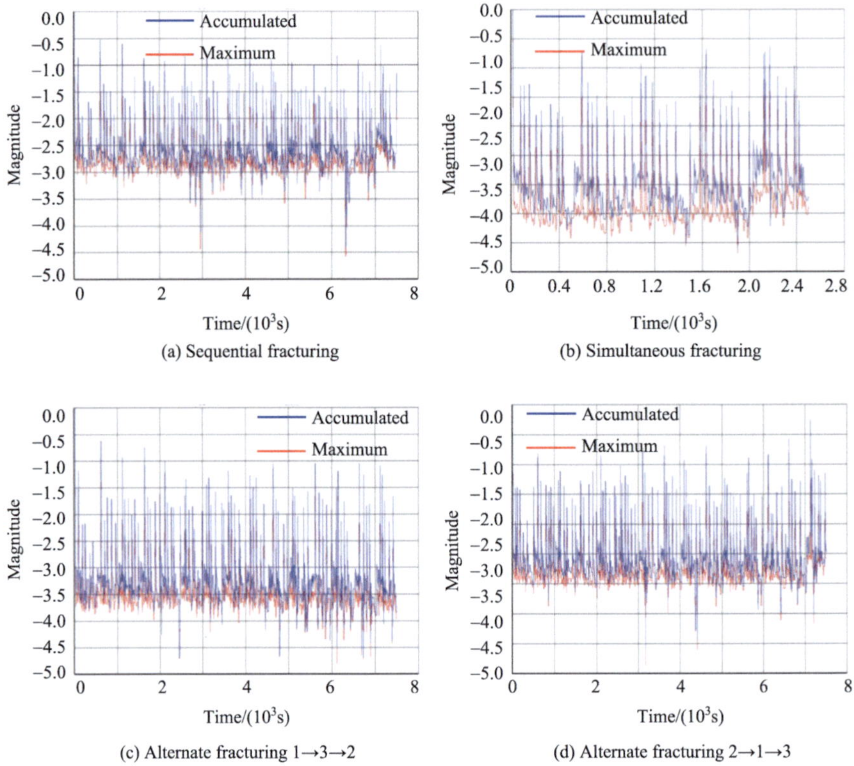

Fig. 6.15 Maximum magnitude and cumulative magnitude change with time under different initiation sequences

(1) In multi-well hydrofracturing, long fractures are connected to other perforation clusters, resulting in fracture hits between wells, affecting the propagation of other fractures which further affects the complexity of the fracture network, It also adversely affects the production of oil and gas reservoirs. With the distribution of parallel perforation clusters in multiple wells, long fractures connect with other perforation clusters in adjacent wells, resulting in fracture hits. A multi-well crossed perforation cluster distribution can effectively avoid the connection between long and adjacent fractures and can also improve the propagation of a single fracture, forming a complex fracture network. The multi-well crossed perforation cluster distribution can significantly reduce the stress disturbance between wells such that the degree of fracture deflection is smaller than that under multi-well parallel perforation clusters.

(2) The variation trend of fracture propagation and stress shadow under different well spacings of the crossed perforation cluster is similar to that of the parallel perforation cluster. With a decrease in well spacing, the shear stress shadow

area formed by the propagation of different wells increases, so that the formation of fracture deflection also increases. The total fracture length increases as well spacing increases. The maximum total fracture volume is at 75 m. As the well spacing decreases, the number of microseismic events decreases, and the fluctuation range of the maximum and cumulative magnitude also decreases.

(3) The variation trends of the fracture propagation and stress shadow under different fracturing sequences of the crossed perforation cluster is similar to those under the parallel perforation cluster. The total fracture length under simultaneous fracturing is the minimum, and the fracture length under alternate fracturing 2 → 1 → 3 (Case IX) is the maximum. The total fracture volume under alternate fracturing 1 → 3 → 2 (Case VIII) is the minimum, while the fracture volume under simultaneous fracturing is the maximum. For the microseismic events produced by different fracturing sequences, the fluctuation amplitude of simultaneous fracturing is the smallest, and the microseismic events are also the smallest. Sequential fracturing has the largest fluctuation range and the largest number of microseismic events. The fluctuation amplitude of alternate fracturing 1 → 3 → 2 and 2 → 1 → 3 (Cases VIII and IX) are moderate, and the number of microseismic events are lower than that of sequential fracturing and higher than that of simultaneous fracturing.

In this study, engineering-scale numerical models were used to investigate the wells connection and connected long hydraulic fractures induced by multi-well hydrofracturing using crossed perforation clusters. To study the spatial expansion behaviour of hydraulic fractures in multiple wells, a three-dimensional numerical model needs to be established to obtain the spatial propagation and interaction behaviours of hydraulic fractures, which will be carried out in future research.

References

Daneshy, A., & Pomeroy, M. (2012). In-situ measurement of fracturing parameters from communication between horizontal wells. *SPE Annual Technical Conference and Exhibition, 6*, 4652–4661.

Daneshy, A., Touchet, C., Hoffman, F., & McKown, M. (2015). Field determination of fracture propagation mode using downhole pressure data. In *SPE Hydraulic Fracturing Technology Conference*, SPE-173345-MS, pp. 303–318.

Duan, K., Li, Y., & Yang, W. (2021). Discrete element method simulation of the growth and efficiency of multiple hydraulic fractures simultaneously-induced from two horizontal wells. *Geomechanics and Geophysics for Geo-Energy and Geo-Resources, 7*(1), 1–20.

Esquivel, R., & Blasingame, T. A. (2017). Optimizing the development of the Haynesville shale-lessons learned from well-to-well hydraulic fracture interference. In *SPE/AAPG/SEG Unconventional Resources Technology Conference*, URTEC-2670079-MS.

Guo, J., Xie, H., Wang, G., Dang, Y., & Li, Y. (2016a). Well pattern and space deployment of horizontal well to prevent fracturing channeling in fractured reservoir. In *6th International Conference on Mechatronics, Materials, Biotechnology and Environment (ICMMBE 2016)*, Vol. 35, pp. 574–579.

Guo, J., Dang, Y., Wang, G., & Li, Y. (2016b). Analysis of the geological characteristics of horizontal well fracturing channeling in Honghe oilfield. In *2016 6th International Conference on Management, Education, Information and Control (MEICI 2016)*, Vol. 135, pp. 207–211.

He, Y., Yang, Z., Li, X., & Song, R. (2020). Numerical simulation study on three-dimensional fracture propagation of synchronous fracturing. *Energy Science & Engineering, 8*(4), 944–958.

Izadi, G., Gaither, M., Cruz, L., Baba, C., Moos, D., & Fu, P. (2015). Fully 3D hydraulic fracturing model: optimizing sequence fracture stimulation in horizontal wells. In *49th US Rock Mechanics/Geomechanics Symposium*, Vol. 3, pp. 1785–1792.

Jo, H. (2012). Optimizing fracture spacing to induce complex fractures in a hydraulically fractured horizontal wellbore. *SPE Americas Unconventional Resources Conference, 2012*, 239–252.

Ju, Y., Li, Y., Wang, Y., & Yang, Y. (2020). Stress shadow effects and microseismic events during hydrofracturing of multiple vertical wells in tight reservoirs: a three-dimensional numerical model. *Journal of Natural Gas Science and Engineering, 84*, 103684.

King, G. E., Rainbolt, M. F., & Swanson, C. (2017). Frac hit induced production losses: evaluating root causes, damage location, possible prevention methods and success of remedial treatments. In *SPE Annual Technical Conference and Exhibition*, SPE-187192-MS.

Li, S., & Zhang, D. (2018). A fully coupled model for hydraulic-fracture growth during multiwell-fracturing treatments: Enhancing fracture complexity. *SPE Production & Operations, 33*(2), 235–250.

Lin, Y., Hu, A., Chen, F., Xiong, P., & Yao, C. (2013). Horizontal well fracturing channeling cause analysis and channeling prevention countermeasures in Honghe oilfield. *Reservoir Evaluation and Development, 3*(4), 56–61.

Liu, X., Rasouli, V., Guo, T., Qu, Z., Sun, Y., & Damjanac, B. (2020). Numerical simulation of stress shadow in multiple cluster hydraulic fracturing in horizontal wells based on lattice modelling. *Engineering Fracture Mechanics*, 238, 107278.

Marongiu-Porcu, M., Lee, D., Shan, D., & Morales, A. (2016). Advanced modeling of interwell-fracturing interference: An eagle ford shale-oil study. *SPE Journal, 21*(5), 1567–1582.

Qiu, F., Porcu, M. M., Xu, J., Malpani, R., Pankaj, P., & Pope, T. L. (2015). Simulation study of zipper fracturing using an unconventional fracture model. In *SPE/CSUR Unconventional Resources Conference*, SPE-175980-MS.

Wang, Y. L. (2020). Adaptive finite element-discrete element analysis for stratal movement and microseismic behaviours induced by multistage propagation of three-dimensional multiple hydraulic fractures. *Engineering Computations, 38*(6), 2781–2809.

Wang, Y. L., Ju, Y., & Yang, Y. M. (2018). Adaptive finite element-discrete element analysis for microseismic modelling of hydraulic fracture propagation of perforation in horizontal well considering pre-existing fractures. *Shock and Vibration, 2018.*

Wang, Y. L., Ju, Y., Chen, J. L., & Song, J. X. (2019). Adaptive finite element-discrete element analysis for the multistage supercritical CO_2 fracturing of horizontal wells in tight reservoirs considering pre-existing fractures and thermal-hydro-mechanical coupling. *Journal of Natural Gas Science and Engineering, 61*, 251–269.

Wang, T., Tian, S., Zhang, W., Ren, W., & Li, G. (2020). Production model of a fractured horizontal well in shale gas reservoirs. *Energy & Fuels, 35*(1), 493–500.

Wang, Y. L., & Liu, N. N. (2022). Dynamic propagation and shear stress disturbance of multiple hydraulic fractures: Numerical cases study via multi-well hydrofracturing model with varying adjacent spacings. *Energies, 15*(13), 4621.

Yu, W., & Sepehrnoori, K. (2013). Optimization of multiple hydraulically fractured horizontal wells in unconventional gas reservoirs. *Journal of Petroleum Engineering, 2013*, 151898.

Chapter 7
Deflection of Fracture Networks and Gas Production in Multi-well Hydrofracturing Utilizing Parallel and Crossed Perforation Clusters

7.1 Introduction

Hydrofracturing is the main technology for unconventional oil and gas extraction (Aloulou & Zaretskaya, 2016; Nagel et al., 2013; Roussel & Sharma, 2011), and the complexity of the hydraulic fracture network is the key to improving unconventional oil and gas production (Wang & Zhang, 2022; Wang et al., 2020, 2022). Multi-well hydrofracturing can form complex fracture network and improve reservoir permeability; compared with single-well hydrofracturing, the fracture interaction between multi-well hydrofracturing is more complex, and the effects of well spacing and fracturing scenarios on fracture network propagation should be considered (Kumar & Ghassemi, 2016; Wang & Liu, 2021). Figure 7.1 is the schematic of fracture networks and gas production in multi-well hydrofracturing utilizing multiple perforation clusters (*a*: perforation cluster spacing; *b*: well spacing). The fractures of multi-well hydrofracturing disturb the newly generated fractures in subsequent wells (Nagel et al., 2014); multi-well hydrofracturing will lead to simultaneous disturbance between fractures in different wells, increase the possibility and range of fracture deflection, and greatly increase the complexity of fracture network (He et al., 2017; Manriquez, 2018). The optimization of hydraulic fractures and the evaluation of gas production require accurate control of fracture deflection.

Well spacing in multi-well hydrofracturing affects fracture network propagation, and reasonable well spacing may improve unconventional oil and gas extraction (Bunger et al., 2011). When perforation clusters in adjacent horizontal wells are designed and stimulated, the hydraulic fractures in the subsequent wells will propagate away from the previous well; if the well spacing is too large, the effective transformation volume of the reservoir between wells decreases, and the resources in the untransformed area may remain in the formation forever; once the well spacing is too small, the risk of affecting the development benefit will increase due to the stress disturbance during hydrofracturing. Hence, the adjacent well spacing should

Y. Wang, *Deflection of Fracturing Fracture Network Disturbed by Discontinuity and Multiple Fractures in Rock*, https://doi.org/10.1007/978-981-95-1569-1_7

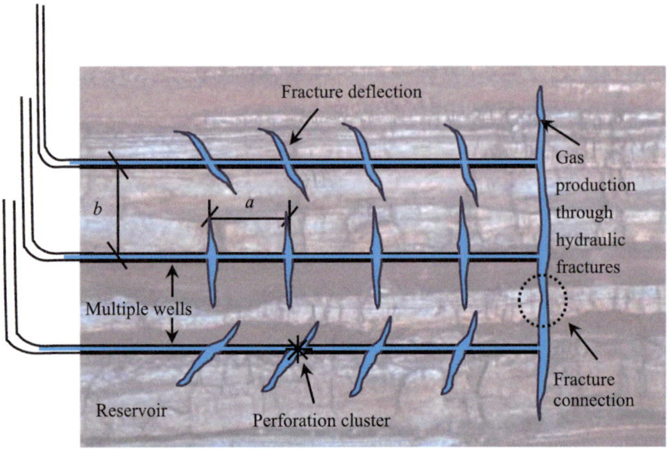

Fig. 7.1 Schematic of fracture networks and gas production in multi-well hydrofracturing utilizing multiple perforation clusters

be optimized by using the induced stress field evolution and disturbance changes during hydrofracturing (Duan et al., 2021).

The fracturing scenarios of single-well hydrofracturing usually include sequential, simultaneous, and alternate fracturing (Wang & Liu, 2021). The fracturing scenarios of multiple horizontal wells also have an important impact on fracture propagation; different fracturing scenarios may form different fracture morphologies, which lead to significantly different gas production rates. The initiation sequence and crossed arrangement of perforation clusters in multiple horizontal wells propose more options for fracturing scenarios (Chen et al., 2018; Yang et al., 2019). To improve the stimulated reservoir volume, the alternate and zipper fracturing (crossed perforation clusters) are developed (Kumar & Ghassemi, 2016). For multi-well zipper hydrofracturing, the conventional parallel perforation clusters distribution is changed to crossed distribution, which has shown advantages in controlling fracture propagation. However, the mechanisms of crossed perforation clusters affecting fracture deflection and gas production are still unclear.

This chapter is organized as follows. Section 7.2 introduces the combined finite element-discrete element method and model considering thermal-hydro-mechanical coupling. Section 7.3 presents the simulation study of deflection of fracture networks in multi-well hydrofracturing utilizing parallel and crossed perforation clusters. Section 7.4 presents the gas production in multi-well hydrofracturing utilizing parallel and crossed perforation clusters. Section 7.5 summarizes the conclusions of this study.

7.2 Combined Finite Element-Discrete Element Method and Model Considering Thermo-Hydro-Mechanical Coupling

7.2.1 Governing Partial Differential Equations

In the process of hydrofracturing of reservoir rock, solid stress field, fluid pressure field and temperature field form a mutual coupling process, as shown in Fig. 7.2 (Wang et al., 2019). The governing equations for solid deformation, fluid flow, and heat transfer are as follows:

(1) Solid Deformation

The governing equation of rock mass matrix deformation is:

$$\boldsymbol{L}^T\left(\boldsymbol{\sigma}' - \alpha \boldsymbol{m} \boldsymbol{p}_l\right) + \rho_B \boldsymbol{g} = 0 \tag{7.1}$$

where $\boldsymbol{L}$ is the differential operator, $\boldsymbol{\sigma}'$ is the effective stress tensor, α is the Biot coefficient, $\boldsymbol{m}$ denotes the identity tensor. $\boldsymbol{p}_l$ is the pore fluid pressure of the rock mass, ρ_B is the saturated bulk density of rock mass, $\boldsymbol{g}$ is the gravity vector.

(2) Fluid Flow

The governing equation of seepage in rock matrix is:

$$\operatorname{div}\left[\frac{k}{u_l}(\nabla p_l - \rho_l \boldsymbol{g})\right] = \left(\frac{\phi}{K_l} + \frac{\alpha - \varphi}{K_s}\right)\frac{\mathrm{d}p_l}{\mathrm{d}t} + \alpha\frac{\mathrm{d}\varepsilon_v}{\mathrm{d}t} \tag{7.2}$$

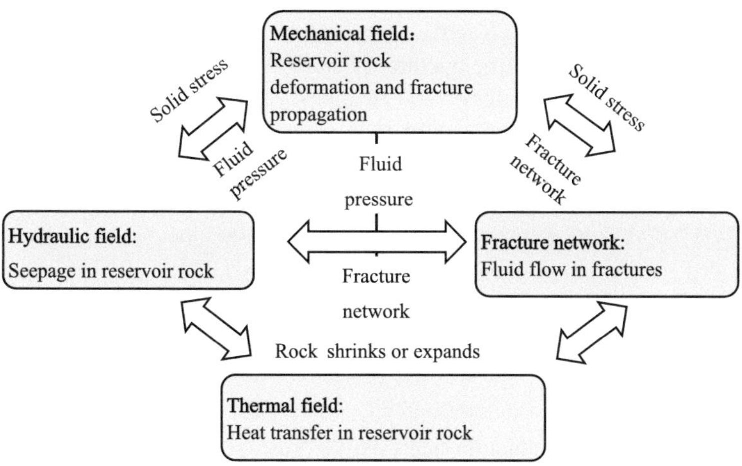

Fig. 7.2 Schematic of thermal-hydro-mechanical coupling and fracture propagation mechanisms

where k is the intrinsic permeability of rock mass, u_l is the viscosity of the pore liquid, ρ_l is the density of the pore liquid, ϕ is the porosity of the porous medium, K_l is the pore fluid stiffness, K_s is the solid skeleton stiffness, ε_v is the volumetric strain of rock mass pore structure and t is the current moment.

The governing equation of fluid flow in the fracture is:

$$\frac{\partial}{\partial x}\left[\frac{k^{fr}}{\mu_n}\left(\nabla p_n - \rho_{fn}\mathbf{g}\right)\right] = S^{fr}\frac{\mathrm{d}p_n}{\mathrm{d}t} + \alpha(\Delta\dot{e}_\varepsilon) \tag{7.3}$$

(3) Heat Transfer

The governing equation of heat transfer between rock matrixes, hydrofracturing fluids is:

$$\mathrm{div}\left[k_b \nabla T_f\right] = \rho_b c_b \frac{\partial T_f}{\partial t} + \rho_f c_f \boldsymbol{q}_f \nabla T_f \tag{7.4}$$

where k_b is the thermal conductivity coefficient, T_f is the fluid temperature, ρ_b is the volume density, c_b is the specific heat coefficient, ρ_f is the fluid density, c_f is the specific heat coefficient of the fluid, and $\boldsymbol{q}_f$ is the Darcy fluid flux.

The heat transfer from the fracture network element from the fluid in the fracture zone to the rock matrix is as follows:

$$q_c^1 = \alpha_c(T_N)\left(T_N - T_f^1\right) \tag{7.5}$$

$$q_c^2 = \alpha_c(T_N)(T_N - T_f^2) \tag{7.6}$$

where q_c^1 and q_c^2 are the contact heat flows between the fractures network and the formation nodes, respectively, α_c is the contact thermal conductivity, T_N is the temperature value of the node within the fracture, T_f^1 and T_f^2 are the corresponding formation node temperatures.

Heat transfer through the reservoir rock causes the rock to shrink or expand, causing stress changes. The volume change depends on the linear thermal expansion coefficient:

$$\Delta V/V = \alpha_T \Delta T \tag{7.7}$$

where V is the initial volume, ΔV is the incremental volume, ΔT is the incremental temperature, and α_T is the linear thermal-expansion coefficient of the rock matrix.

(4) Gas Production

For gas recovery and production analysis, the gas seepage equation and the gas network equation, which combines mass conservation with Darcy's law, are given by:

$$\mathrm{div}\left[\frac{k(p_g)}{\mu_g}\nabla p_g - \rho_g g\right] = \left[\phi\frac{\partial \rho_g}{\partial p_g} + (\rho_g - q)\frac{\partial \phi}{\partial p_g} + (1-\phi)\phi\frac{\partial q}{\partial p_g}\right]\frac{\partial p_g}{\partial t} \tag{7.8}$$

$$\frac{\partial}{\partial x}\left[\frac{K^{fr}}{\mu_g}(\nabla p_g - \rho_g \mathbf{g})\right] = \phi\left(C_g - \frac{\rho_g}{Z}\frac{\partial \rho_g}{\partial Z}\right)\frac{\partial \rho_g}{\partial t} \tag{7.9}$$

where μ_g is the viscosity of the pore gas; p_g is the pore gas pressure; ρ_g is the density of the pore gas; ϕ is the porosity of the porous media; q is the mass of adsorbed gas per unit volume; C_g is the gas compressibility; and Z is the gas compressibility factor.

7.2.2 Numerical Models of Multi-well Hydrofracturing

In this study, two different perforation cluster distribution patterns of multi-well hydrofracturing are studied, which include parallel and crossed perforation clusters as shown in Figs. 7.3 and 7.4, respectively. There are three horizontal wells (denoted as Well 1, Well 2, and Well 3), and five perforation clusters (numbered 1–5 in sequence) were set for each horizontal well; a is the perforation cluster spacing and b is the well spacing; the blue box area in the diagram is the mesh refinement domain. According to the actual reservoir project, the physical parameters of numerical models of multi-well hydrofracturing are shown in Table 7.1.

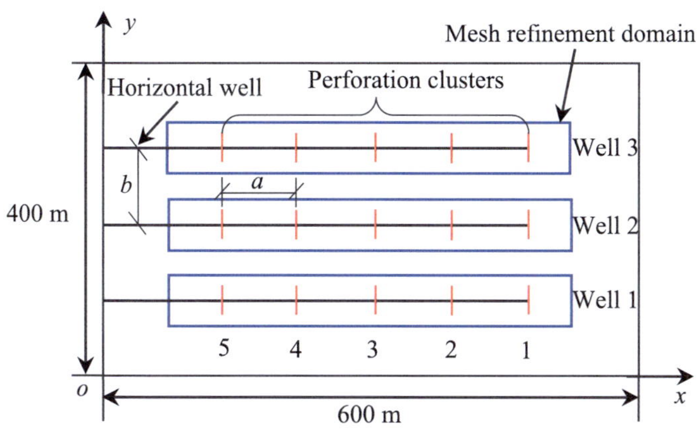

Fig. 7.3 Geometric model of multi-well multistage hydrofracturing utilizing parallel perforation clusters

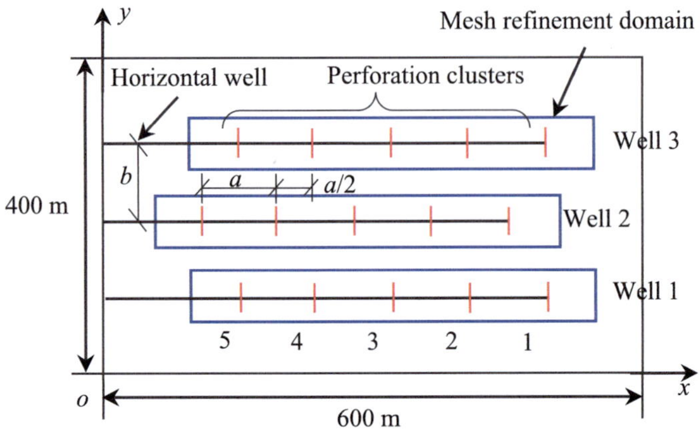

Fig. 7.4 Geometric model of multi-well multistage hydrofracturing utilizing crossed perforation clusters

Table 7.1 Physical parameters of numerical models of multi-well hydrofracturing

Parameters	Value
Horizontal minimum in-situ stress (x-direction) S_h/MPa	40
Horizontal maximum in-situ stress (y-direction) S_H/MPa	44
Fluid injection rate Q/(m^3/s)	0.5
Pore pressure p_s/MPa	10
Biot coefficient α	0.75
Elastic modulus E/GPa	31
Poisson's ratio ν	0.22
Connection k/nD	50
Porosity ϕ	0.05
Kinematic viscosity coefficient μ_n/(Pa · s)	1.67×10^{-3}
Fracture fluid bulk modulus K_f^{fr}/MPa	2000
Tensile strength σ_t/MPa	5.26
Fracture energy G_f/(N · m)	165

7.2.2.1 Fracturing Scenarios and Well Spacings for Parallel Perforation Clusters

To investigate the fracturing scenarios and perforation spacings of parallel perforation clusters, some cases are set as shown in Table 7.2:

(a) The perforation cluster spacing a of each well is 75 m, and the perforation clusters in each well use sequential fracturing ($1 \rightarrow 2 \rightarrow 3 \rightarrow 4 \rightarrow 5$).
(b) Three well spacings are set as $b = 50$, 75, and 100 m.

Table 7.2 Fracturing scenarios and well spacings for parallel perforation clusters

Cases	Fracturing scenarios	b/m
1-I	Sequential	50
1-II		75
1-III		100
1-IV	Alternate 1-3-2	75
1-V	Alternate 2-1-3	75
1-VI	Simultaneous	75

(c) Four fracturing scenarios are set: sequential fracturing (Well 1 → Well 2 → Well 3), alternate fracturing 1-3-2 (Well 1 → Well 3 → Well 2), alternate fracturing 2-1-3 (Well 2 → Well 1 → Well 3), and simultaneous fracturing.

7.2.2.2 Fracturing Scenarios and Well Spacings for Crossed Perforation Clusters

To investigate the fracturing scenarios and perforation spacings of crossed perforation clusters, some cases are set as shown in Table 7.3:

(a) The crossed perforation clusters distribution is that Well 2 is offset by $a/2$ from Well 1 and 3, resulting in a crossed distribution of perforation clusters on each well.
(b) The perforation cluster spacing a of each well is 75 m, and the perforation clusters in each well use sequential fracturing (1 → 2 → 3 → 4 → 5).
(c) Three well spacings are set as $b = 25$, 35, 50, 75, and 100 m.
(d) Four fracturing scenarios are set: sequential fracturing (Well 1 → Well 2 → Well 3), alternate fracturing 1-3-2 (Well 1 → Well 3 → Well 2), alternate fracturing 2-1-3 (Well 2 → Well 1 → Well 3), and simultaneous fracturing.

Table 7.3 Fracturing scenarios and well spacings for crossed perforation clusters

Cases	Fracturing scenarios	b/m
2-I	Sequential	25
2-II		35
2-III		50
2-IV		75
2-V		100
2-VI	Alternate 1-3-2	25
2-VII	Alternate 2-1-3	
2-VIII	Simultaneous	
2-IX	Alternate 1-3-2	75
2-X	Alternate 2-1-3	
2-XI	Simultaneous	

7.3 Deflection of Fracture Networks in Multi-well Hydrofracturing Utilizing Parallel and Crossed Perforation Clusters

7.3.1 Fracture Deflection in Multi-well Hydrofracturing Utilizing Parallel Perforation Clusters

7.3.1.1 Fracturing Scenarios of Multiple Wells

Through numerical computation, the deflection of fracture networks in multi-well hydrofracturing utilizing parallel perforation clusters are obtained. Figure 7.5 shows the fracture network morphology and gas pressure distribution under different fracturing scenarios of parallel perforation clusters with 75 m of well spacing, and the low-pressure zone is formed around the fracture network, which is conducive to tight unconventional gas migration into the fracture. In Fig. 7.5a, in the process of sequential fracturing, the stress disturbance between the fractures of the three wells is accumulated. Except for the straight propagation of fracture 1 in well 1, the other fractures are deflected or connected by the stress disturbance caused by fractures of adjacent wells. Figure 7.5b shows the fracture networks and gas pressure distribution in alternate fracturing 1-3-2, and the fractures stimulated from the parallel perforation clusters are connected. In the local domains around the adjacent wells, the superposition and reduction of shear stresses occur and these stress disturbances deflect the fractures (Wang & Liu, 2021). Due to the alternate fracturing scheme, the enlargement of relative well spacing weakened the disturbance of stress and the deflection of fractures. Well 1 and 3 have deflected the fractures on the far side away from Well 2, while other fractures almost straightly propagate. The results of alternate fracturing 2-1-3 are shown in Fig. 7.5c, when the fractures in Well 1 and Well 3 propagate, due to the stress disturbance caused by Well 2, the fractures 4 and 5 in Well 1 and Well 3 deflect, and fractures 1 and 2 in each well are connected. As shown in Fig. 7.5d, the fracture 1 in the three wells in simultaneous fracturing are connected, and the nearly symmetrical shear stress at the fracture tip makes the fractures propagate straightly; with the initiation of subsequent fractures, the stress disturbance caused by fracture propagation gradually accumulates, resulting in the aggravation of the deflection of subsequent fractures; Well 1 and Well 3 are located on the outside, and the fractures on them are more likely to deflect and propagate outwards.

7.3.1.2 Well Spacing Between Multiple Wells

In Fig. 7.6, the fracture network morphology and gas pressure distribution under different well spacings of parallel perforation clusters are shown. From 25 to 50 m and

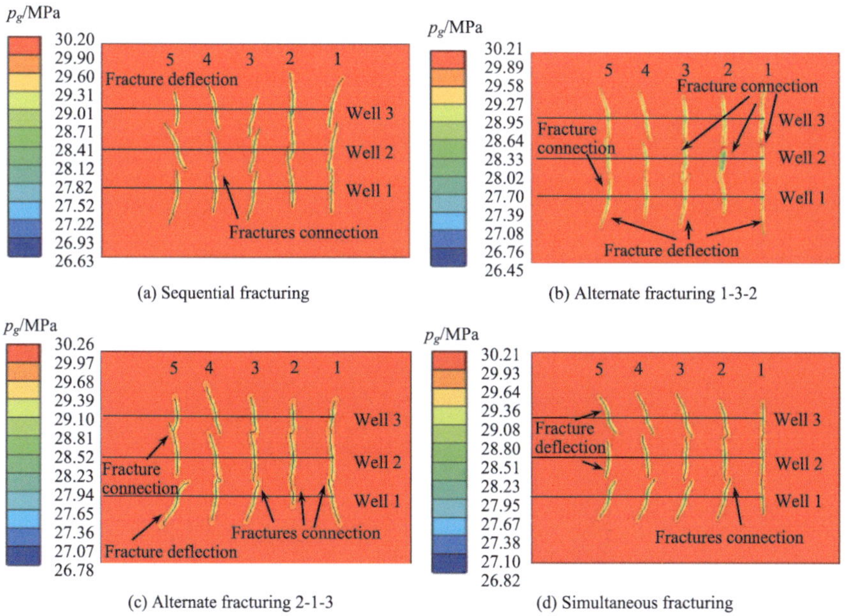

Fig. 7.5 Fracture network morphology and gas pressure distribution under different fracturing scenarios of parallel perforation clusters with 75 m of well spacing

then to 100 m of well spacings, the number of connected or crossed fractures gradually decreases, and the stress disturbance between wells gradually decreases. When the single well is stimulated sequentially, the first fracture propagates straightly, and the subsequent fractures have the characteristics of outward deflection. As the well spacing increases, the fracture deflection also weakens, and all the fractures propagate almost straightly when the well spacing reaches 100 m.

7.3.2 *Fracture Deflection in Multi-well Hydrofracturing Utilizing Crossed Perforation Clusters*

7.3.2.1 Fracturing Scenarios of Multiple Wells

Through numerical computation, the deflection of fracture networks in multi-well hydrofracturing utilizing crossed perforation clusters are obtained. Figure 7.7 shows the fracture propagation and gas pressure distribution under different fracturing scenarios with 75 m well spacing. It can be seen that compared to the parallel perforation clusters scheme as shown in Fig. 7.5, the number of connected frac-

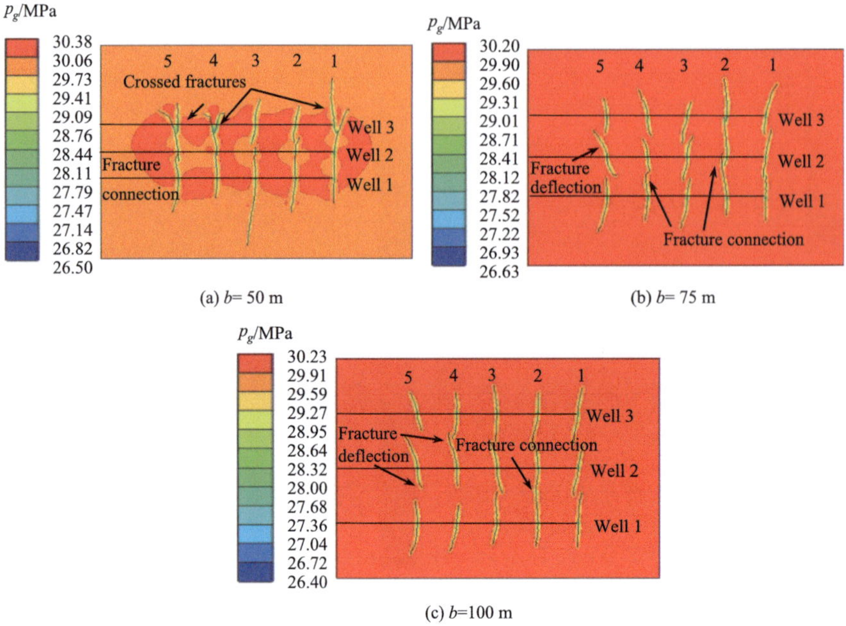

Fig. 7.6 Fracture network morphology and gas pressure distribution under different well spacings of parallel perforation clusters

tures in each horizontal well is reduced, and the deflection of each fracture has also been greatly reduced. This fracturing scheme of crossed perforation clusters under all fracturing scenarios can form a relatively dispersed fracture network on the reservoir. In addition, compared to sequential fracturing (Fig. 7.7a) and simultaneous fracturing (Fig. 7.7d), there is no significant difference in reducing the disturbance of fractures between the two alternate fracturing scenarios (Figs. 7.7b and c), indicating that the fracturing scheme of crossed perforation clusters plays a more dominant role.

Figure 7.8 shows the fracture propagation and gas pressure distribution under different fracturing scenarios with 25 m well spacing. Compared with the fracture propagation of different fracturing scenarios under the well spacing of 75 m (Fig. 7.7), many fractures stimulated in each horizontal well are connected; reducing the well spacing will aggravate the stress disturbance between wells during fractures propagating, and the deflection of each fracture has been enlarged. Special emphasis should be placed here on simultaneous fracturing (Fig. 7.8d), except for the fracture 1 being connected, all other fractures are deflected and avoid the connection between each other.

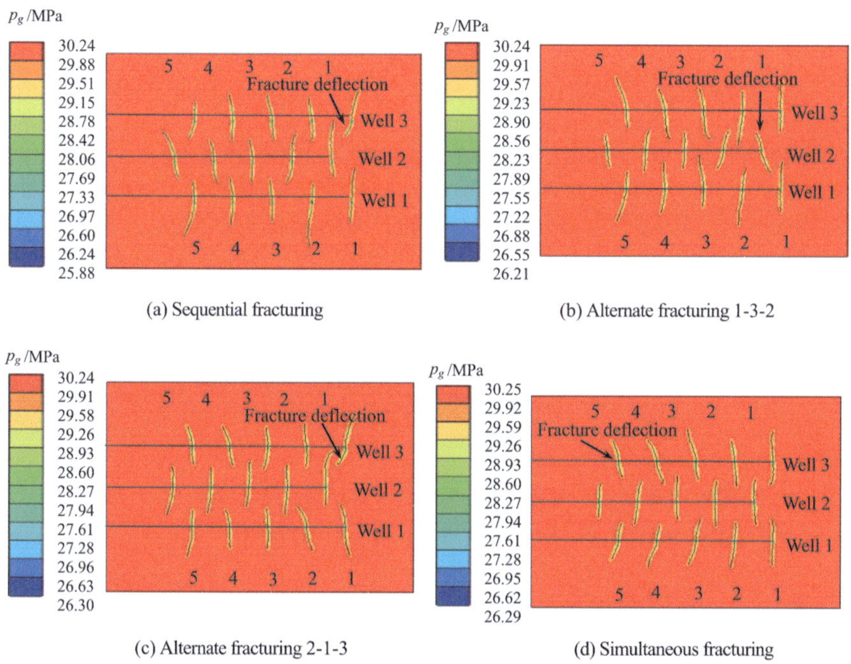

Fig. 7.7 Fracture network morphology and gas pressure distribution under different fracturing scenarios with 75 m of well spacing

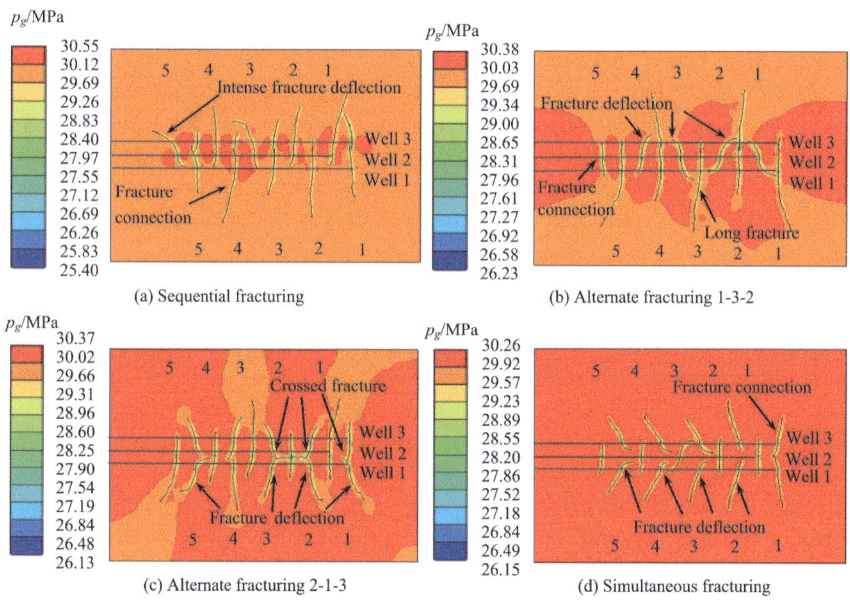

Fig. 7.8 Fracture network morphology and gas pressure distribution under different fracturing scenarios with 25 m of well spacing

7.3.2.2 Well Spacing Between Multiple Wells

In Fig. 7.9, the fracture propagation morphology and gas pressure distribution under different well spacings of crossed perforation clusters are shown. From 25 to 100 m of well spacings, the number of connected fractures quickly decreases, due to the geometric relationship of crossed perforation clusters. When the single well is stimulated sequentially, the first fracture propagates straightly, and the subsequent fractures have the characteristics of outward deflection. As the well spacing increases, the fracture deflection also weakens, and all the fractures almost propagate straightly When the well spacing is 75 m. The above behaviors are similar to the results of parallel perforation clusters, but the crossed perforation clusters can weaken the disturbance between wells, making it easier for fractures to propagate straightly.

7.4 Gas Production in Multi-well Hydrofracturing Utilizing Parallel and Crossed Perforation Clusters

7.4.1 Gas Production in Multi-well Hydrofracturing Utilizing Parallel Perforation Clusters

7.4.1.1 Fracturing Scenarios of Multiple Wells

In this section, the gas production under different fracturing scenarios of 75 m of well spacing in parallel perforation clusters is selected for comparison. As shown in Fig. 7.10 (local figure), some low-pressure zones are formed near the fracturing fractures, which drive the gas flow in the tight reservoir and gas recovery to the fracturing fracture network. Compared with sequential fracturing, the deflection of simultaneous fracturing is larger than that of sequential fracturing; the deflected fractures avoid crossing, which leads to the occurrence of fractures in more zones of the reservoir.

Combined with the gas production volume V_g, fracture length L and fracture volume V of different fracturing scenarios are provided in Table 7.4. In order to intuitively compare the gas production under different fracturing scenarios, Fig. 7.11 shows the gas production results under multiple fracturing scenarios. The fracture length and volume of sequential fracturing is longer than that of simultaneous fracturing, however, the gas production of sequential fracturing (759030.32 m^3) is smaller than that of simultaneous fracturing (781907.50 m^3); the deflected fractures in simultaneous fracturing avoid crossing, which leads to the occurrence of fractures in more zones of the reservoir and is beneficial for the migration and production of tight unconventional gas in the reservoir.

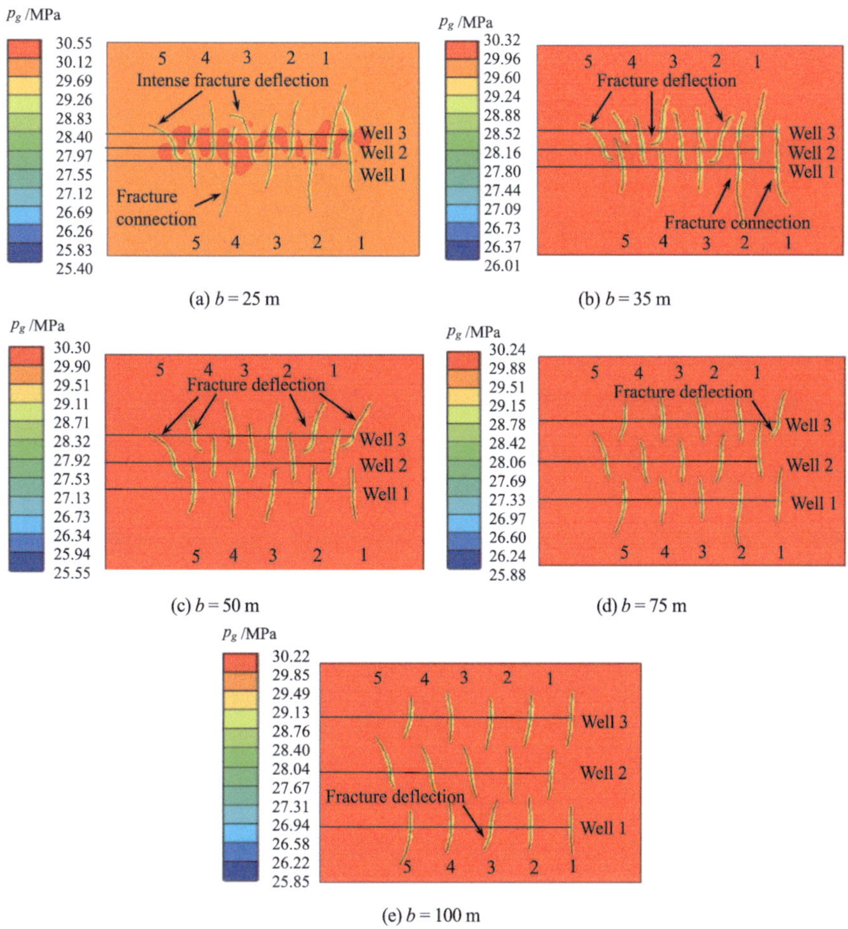

Fig. 7.9 Fracture network morphology and gas pressure distribution under different well spacings of crossed perforation clusters

7.4.1.2 Well Spacing Between Multiple Wells

In this section, the gas production of sequential fracturing under three well spacings (50, 75, and 100 m) in the parallel perforation clusters model is compared to explore the effects of well spacing changes on gas production under sequential fracturing. Figure 7.12 (local figure) shows the fracture network morphology and gas pressure distribution under sequential fracturing with 50 and 100 m of well spacings. When the well spacing is 50 m (Fig. 7.12a), the degree of fracture deflection is large, and there are more crossed and connected fractures; when the well spacing is 100 m

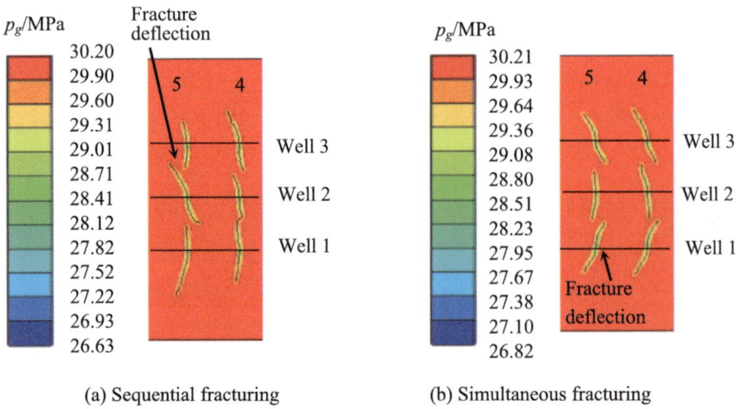

Fig. 7.10 Fracture network morphology and gas pressure distribution under simultaneous fracturing and sequential fracturing with 75 m of well spacing

Table 7.4 Quantitative results of parallel perforation clusters cases under various fracturing scenarios

Cases	Fracturing scenarios	V_g/m^3	L/m	V/m^3
1-II	Sequential	759030.32	1350.35	36.38
1-IV	Alternate 1-3-2	766453.87	1364.25	39.85
1-V	Alternate 2-1-3	777252.52	1232.60	34.20
1-VI	Simultaneous	781907.50	1177.74	30.89

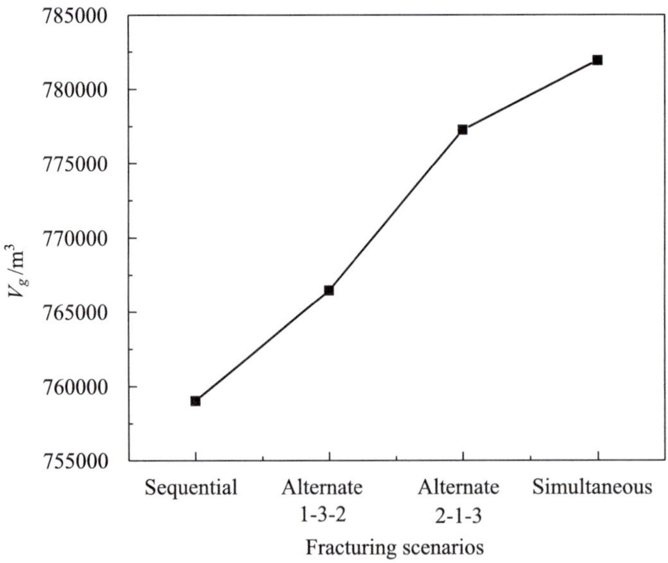

Fig. 7.11 Gas production under multiple fracturing scenarios with 75 m of well spacing

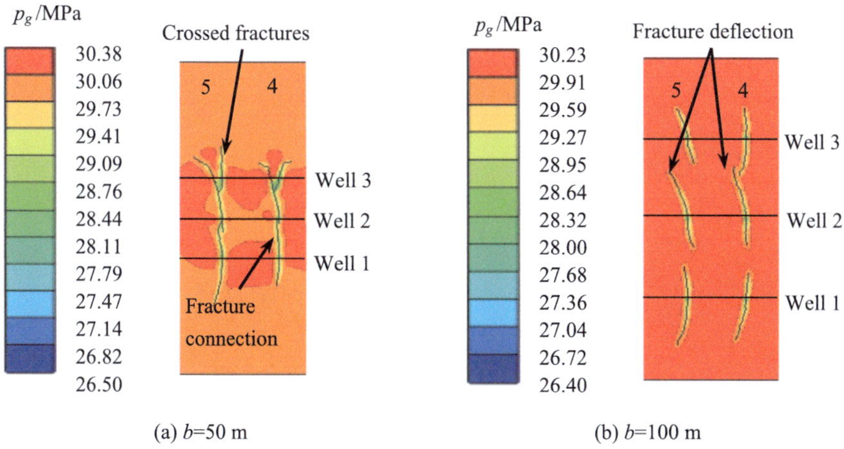

Fig. 7.12 Fracture network morphology and gas pressure distribution under sequential fracturing with 50 m and 100 m of well spacings

(Fig. 7.12b), the degree of fracture deflection decreases, and the connected fractures are greatly reduced.

Table 7.5 lists the quantitative results (V_g, L, V) of parallel perforation clusters cases with different well spacings (50, 75, and 100 m). In order to intuitively compare the gas production changes with different well spacings, Fig. 7.13 shows the gas production results with different well spacings. When the well spacing is 100 m, the gas production is the largest, and the fracture length and volume are also the largest. Under sequential fracturing, the gas production increases with the increase of well spacing. As the spacing between wells increases, the disturbance of fractures stimulated by each well decreases, and the stimulated volume of reservoir by fractures increases, which is conducive to the recovery and production of tight unconventional gas.

Table 7.5 Quantitative results of sequential fracturing utilizing of parallel perforation clusters with different well spacings

Cases	b/m	V_g/m^3	L/m	V/m^3
1-I	50	755117.22	1322.56	36.44
1-II	75	759030.32	1350.35	36.38
1-III	100	765441.18	1450.68	38.33

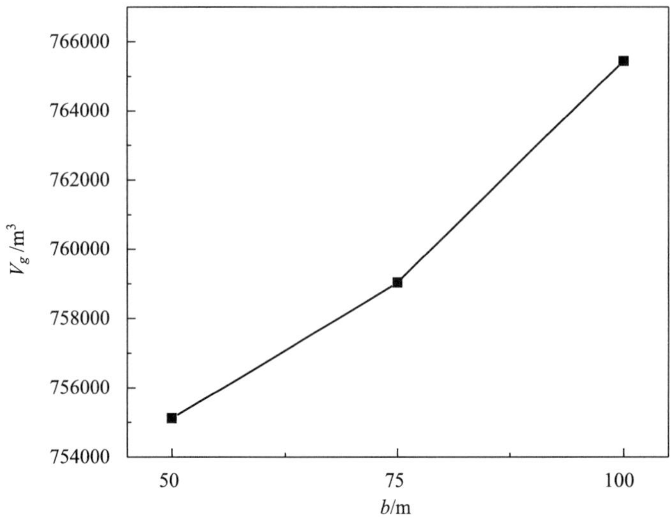

Fig. 7.13 Gas production under different well spacings in sequential fracturing

7.4.2 Gas Production in Multi-well Hydrofracturing Utilizing Crossed Perforation Clusters

7.4.2.1 Fracturing Scenarios of Multiple Wells

In this section, the gas productions under different fracturing scenarios with well spacings of 25 m and 75 m in the cross perforation cluster are selected for comparison. As shown in Fig. 7.14 (local figure), the fractures under sequential fracturing.

Figure 7.14a are slightly deflected with connection, while in the case of alternate fracturing 1-3-2 (Fig. 7.14b), the fractures in the wells on both sides meet and then deflect; compared with sequential fracturing, the deflection of alternate fracturing is larger than that of sequential fracturing; the deflected fractures in sequential fracturing avoid crossing, which leads to the occurrence of fractures in more zones of the reservoir. As shown in Fig. 7.15, when the spacing between wells increases to 75 m, there is still deflection in the fractures, but there will be no further connection between the fractures.

Table 7.6 lists the quantitative results (V_g, L, V) of crossed perforation clusters cases under different fracturing scenarios with 25 and 75 m of well spacings. It can be seen that compared to the parallel perforation clusters scheme as shown in Table 7.4. The crossed perforation clusters scheme can provide more gas production V_g and larger fracture length L. In order to intuitively compare the gas production of crossed perforation clusters cases under different fracturing scenarios, Fig. 7.16 shows the gas production results under multiple fracturing scenarios. For 25 m of well spacing, the alternate fracturing (Alternate 2-1-3) between multiple wells can

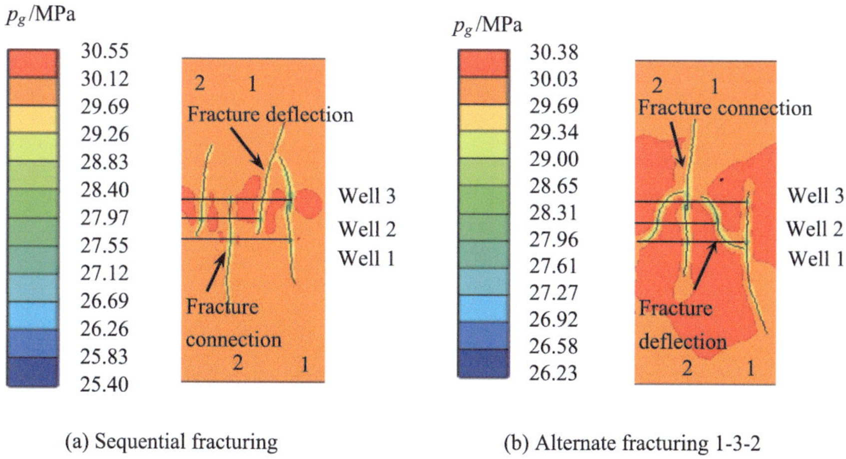

Fig. 7.14 Fracture network morphology and gas pressure distribution under alternate fracturing 1-3-2 and sequential fracturing with 25 m of well spacing

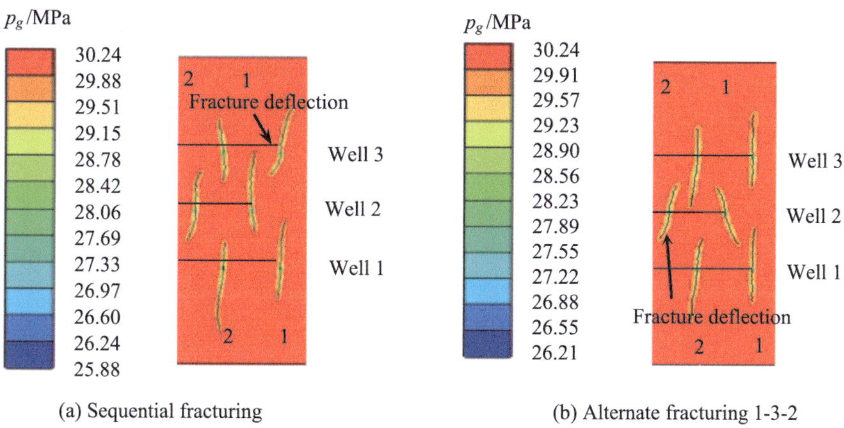

Fig. 7.15 Fracture network morphology and gas pressure distribution under alternate fracturing 1-3-2 and sequential fracturing with 75 m of well spacing

form longer fractures and higher gas production (759030.32 m^3), as it reduces disturbance between wells; the simultaneous fracturing between multiple wells can form shorter fractures and lower gas production (816745.49 m^3), as simultaneous fracturing increases disturbance between wells. For 75 m of well spacing, due to the distribution of hydraulic fractures in larger reservoirs caused by multi-well hydrofracturing, the gas production in this situation is significantly increased compared to when the well spacing is 25 m; the alternate fracturing is still the optimized choice for increasing gas production in various fracturing scenarios.

Table 7.6 Quantitative results of crossed perforation clusters cases under different fracturing scenarios

Cases	Fracturing scenarios	*b*/m	V_g/m^3	*L*/m	*V*/m^3
2-I	Sequential	25	843480.31	1399.43	42.85
2-VI	Alternate 1-3-2		831382.81	1393.80	37.33
2-VII	Alternate 2-1-3		850113.10	1530.04	43.90
2-VIII	Simultaneous		816745.49	1252.85	29.84
2-IV	Sequential	75	857463.24	1382.34	39.50
2-IX	Alternate 1-3-2		873274.89	1365.50	40.63
2-X	Alternate 2-1-3		879676.36	1331.86	37.69
2-XI	Simultaneous		854352.15	1268.81	30.16

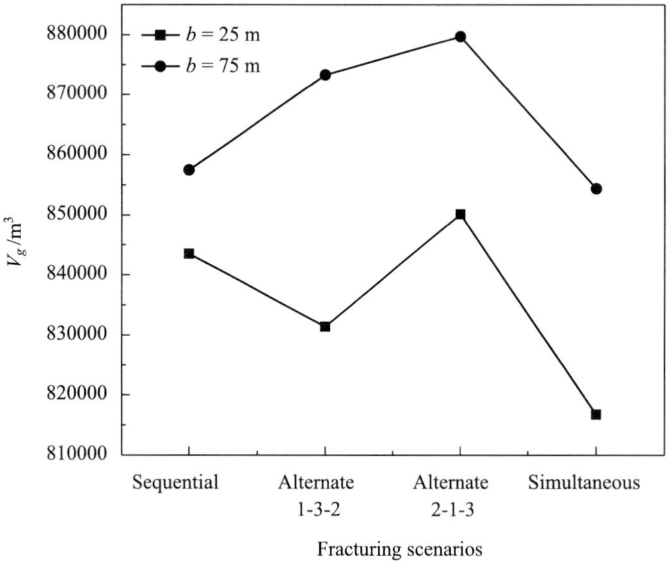

Fig. 7.16 Gas production under different fracturing scenarios

7.4.2.2 Well Spacing Between Multiple Wells

The fracture network morphology and gas pressure distribution under sequential fracturing utilizing crossed perforation clusters with 25 and 35 m of well spacings are compared to detect the inference of well spacing between multiple wells, as shown in Fig. 7.17 (local figure). When the well spacing is 25 m, the fractures on both sides of the adjacent wells are connected; when the well spacing is 35 m, the number of connected fractures decreases. At small well spacing (35 m), the fracturing scheme of crossed perforation clusters may reduce the connection between fractures.

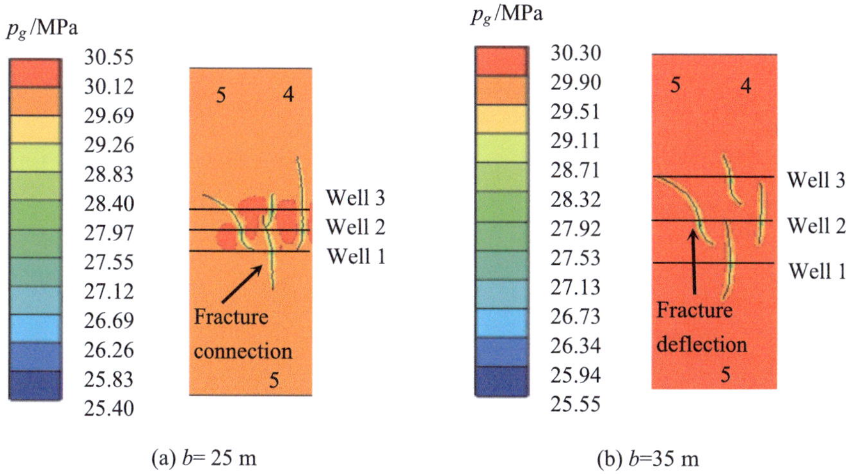

Fig. 7.17 Fracture network morphology and gas pressure distribution under sequential fracturing with 25 and 35 m of well spacings

Table 7.7 lists the quantitative results (V_g, L, V) of crossed perforation clusters cases with different well spacings (25, 35, 50, 75, and 100 m). In order to intuitively compare the gas production changes with different well spacings, Fig. 7.18 shows the gas production results in sequential fracturing with different well spacings. When the well spacing is relatively small, the disturbance of fractures becomes stronger, which results in relatively less gas production as the spacing between wells is 35 m; when the well spacing is 25 m, the fractures are connected, forming relatively long fractures, which is conducive to promoting the gas recovery and production. When the well spacing increases, the disturbance of fractures weakens, resulting in a higher gas production rate as the spacing between wells is 75 m; when the well spacing is 100 m, the fractures are not connected and form independent fractures, which is not conducive to promoting the gas recovery and production.

Table 7.7 Quantitative results of sequential fracturing utilizing of crossed perforation clusters with different well spacings

Cases	b/m	V_g/m^3	L/m	V/m^3
2-I	25	843480.31	1399.43	42.85
2-II	35	828658.42	1470.84	41.14
2-III	50	837161.54	1395.25	34.69
2-IV	75	857463.24	1382.34	39.50
2-V	100	835971.45	1416.02	37.00

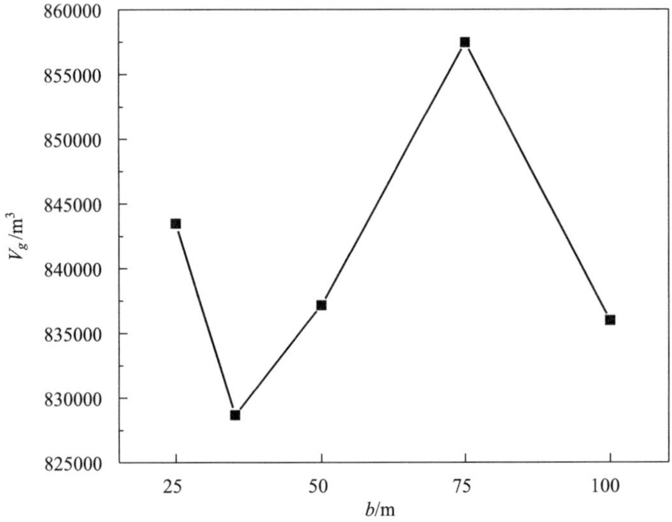

Fig. 7.18 Gas production in sequential fracturing with different well spacings

7.5 Conclusions

The conclusions of this study are as follows:

(1) The numerical models for analyzing the deflection of fracture networks and gas production in multi-well hydrofracturing utilizing parallel and crossed perforation clusters are established. The combined finite element-discrete element method is used, and the coupling effects of solid, fluid, and temperature fields are considered. The numerical cases of multi-well hydrofracturing under varying fracturing scenarios and well spacings and are simulated.

(2) For deflection of fracture networks in multi-well hydrofracturing utilizing parallel perforation clusters, due to the alternate fracturing scheme, the enlargement of relative well spacing weakened the disturbance of stress and the deflection of fractures; as the well spacing increases, the fracture deflection also weakens, and all the fractures propagate almost straightly. Once the crossed perforation clusters scheme is utilized, the number of connected fractures in each horizontal well is reduced, and the deflection of each fracture has also been greatly reduced.

(3) For gas production in multi-well hydrofracturing utilizing parallel perforation clusters, the deflected fractures in simultaneous fracturing avoid crossing, which leads to the occurrence of fractures in more zones of the reservoir and is beneficial for the migration and production of tight unconventional gas in the reservoir; as the spacing between wells increases, the disturbance of fractures stimulated by each well decreases, and the stimulated volume of reservoir by fractures

increases, which is conducive to the recovery and production of tight unconventional gas. Compared to the parallel perforation clusters scheme, the crossed perforation clusters scheme can provide more gas production and larger fracture length.

In this study, the in-plane behaviors of hydraulic fractures between multiple wells were studied, which can provide some references for practical engineering. In the near future study, the spatial effects and disturbances of fracture propagation will be introduced to investigate the propagation and deflection behaviors of fracture in three-dimensional morphology.

References

Aloulou, F., & Zaretskaya, V. (2016). Shale gas production drives world natural gas production growth. *US Energy Information Administration, 218*, 385–395.
Bunger, A. P., Jeffrey, R. G., Kear, J., Zhang, X., & Morgan, M. (2011). Experimental investigation of the interaction among closely spaced hydraulic fractures. In *The 45th U.S Rock Mechanics/ Geomechanics symposium*, American Rock Mechanics Association, ARMA-11-318.
Chen, X., Li, Y., Zhao, J., Xu, W., & Fu, D. (2018). Numerical investigation for simultaneous growth of hydraulic fractures in multiple horizontal wells. *Journal of Natural Gas Science and Engineering, 51*, 44–52.
Duan, K., Li, Y. C., & Yang, W. D. (2021). Discrete element method simulation of the growth and efficiency of multiple hydraulic fractures simultaneously-induced from two horizontal wells. *Geomechanics and Geophysics for Geo-energy and Geo-resources, 7*(3).
He, Q. Y., Suorineni, F. T., Ma, T. H., & Oh, J. (2017). Effect of discontinuity stress shadows on hydraulic fracture re-orientation. *International Journal of Rock Mechanics and Mining Sciences, 91*, 179–194.
Kumar, D., & Ghassemi, A. (2016). A three-dimensional analysis of simultaneous and sequential fracturing of horizontal wells. *Journal of Petroleum Science and Engineering, 146*, 1006–1025.
Manriquez, A. L. (2018). Stress behavior in the near fracture region between adjacent horizontal wells during multistage fracturing using a coupled stress-displacement to hydraulic diffusivity model. *Journal of Petroleum Science and Engineering, 162*, 822–834.
Nagel, N., Zhang, F., Sanchez-Nagel, M., Lee, B., & Agharazi, A. (2013). Stress shadow evaluations for completion design in unconventional plays. In *SPE Unconventional Resources Conference Canada*, SPE-167128-MS.
Nagel, N., Sheibani, F., Lee, B., Agharazi, A., & Zhang, F. (2014). Fully-coupled numerical evaluations of multi-well completion schemes: the critical role of in-situ pressure changes and well configuration. In *SPE Hydraulic Fracturing Technology Conference*, SPE-168581-MS.
Roussel, N. P., & Sharma, M. M. (2011). Optimizing fracture spacing and sequencing in horizontal-well fracturing. *SPE Production & Operations, 26*(2), 173–184.
Wang, Y., & Liu, X. (2021). Stress-dependent unstable dynamic propagation of three dimensional multiple hydraulic fractures with improved fracturing sequences in heterogeneous reservoirs: Numerical cases study via poroelastic effective medium model. *Energy & Fuels, 35*(22), 18543–18562.
Wang, Y., & Zhang, X. (2022). Dual bilinear cohesive zone model-based fluid-driven propagation of multiscale tensile and shear fractures in tight reservoir. *Engineering Computations, 39*(10), 3416–3441.
Wang, Y., Ju, Y., Chen, J., & Song, J. (2019). Adaptive finite element-discrete element analysis for the multistage supercritical CO_2 fracturing and microseismic modelling of horizontal wells

in tight reservoirs considering pre-existing fractures and thermal-hydro-mechanical coupling. *Journal of Natural Gas Science and Engineering, 61*, 251–269.
Wang, T., Tian, S., Zhang, W., Ren, W., & Li, G. (2020). Production model of a fractured horizontal well in shale gas reservoirs. *Energy & Fuels, 35*(7), 493–500.
Wang, Y., Wang, J., & Li, L. (2022). Dynamic propagation behaviors of hydraulic fracture networks considering hydro-mechanical coupling effects in tight oil and gas reservoirs: A multi-thread parallel computation method. *Computers and Geotechnics, 152*, Article 105016.
Yang, Z., He, R., Li, X., Li, Z., Liu, Z., & Lu, Y. (2019). Application of multi-vertical well simultaneous hydraulic fracturing technology for deep coalbed ethane (DCBM) production. *Chemistry and Technology of Fuels and Oils, 55*, 299–309.

Chapter 8
Supercritical-CO_2-driven Intersections of Multi-Well Fracturing Fracture Network and Induced Microseismic Events in Naturally Fractured Reservoir

8.1 Introduction

Supercritical CO_2 (SC-CO_2) offers distinct advantages over conventional aqueous fracturing fluids for enhancing shale production and reservoir preservation (Peng et al., 2017; Ranjith et al., 2019). Figure 8.1 illustrates multi-well SC-CO_2 fracturing systems, including heat transfer, fluid dynamics, and induced microseismicity. SC-CO_2's low viscosity, high permeability, and density promote complex fracture networks conducive to shale gas extraction. Furthermore, its subsurface sequestration provides environmentally sustainable greenhouse gas mitigation (Wang et al., 2012). However, CO_2 exists in gaseous phase at ambient conditions, transitioning to supercritical phase under elevated temperature/pressure (Zhao et al., 2021). This necessitates comprehensive thermal-hydro-mechanical (THM) coupling analysis to maintain critical-state conditions during fracturing operations.

Current methodologies inadequately characterize multi-well SC-CO_2 fracturing mechanisms in naturally fractured reservoirs, presenting persistent challenges including: fracture-network interactions between stimulated and natural fractures, thermal transport phenomena, THM coupling effects, quantitative fracture network diagnostics, induced formation damage, and contact-slip events. Elucidating these unconventional fracturing mechanisms would establish robust predictive frameworks for optimizing hydraulic fracturing operations in hydrocarbon extraction.

Studies demonstrate that temperature fluctuations during SC-CO_2 fracturing induce thermal stresses that promote micro-fractures generation along fracture surfaces (Li et al., 2018, 2023; Liao et al., 2023; Wang et al., 2019). Elevated fracturing fluid temperatures enhance CO_2 diffusivity, generating compressive stresses on fracture walls, while formation cooling induces tensile stresses—both critically influencing fracture initiation and propagation. In fractured reservoirs, thermally-induced tensile stresses increase fracture apertures while reducing injection pressures, with greater fluid-formation temperature differentials amplifying these effects and increasing morphological complexity. This facilitates secondary

Y. Wang, *Deflection of Fracturing Fracture Network Disturbed by Discontinuity and Multiple Fractures in Rock*, https://doi.org/10.1007/978-981-95-1569-1_8

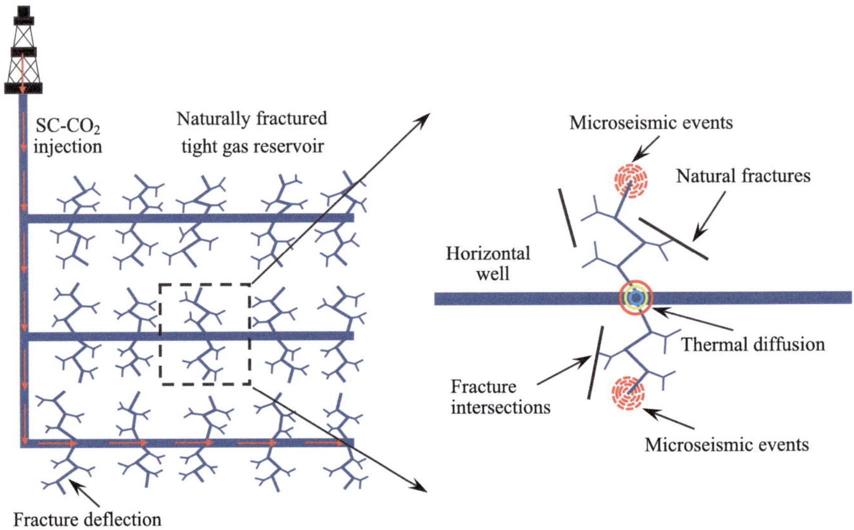

Fig. 8.1 Schematic simulation of supercritical CO_2-driven intersections of multi-well fracturing fracture networks and induced microseismic events in naturally fractured reservoir

fracture propagation orthogonal to primary fractures. Laboratory evidence further confirms that SC-CO_2 fracture pathways respond to reservoir discontinuities through either bedding-plane penetration or arrest, generating extensive micro-fractures networks along slip surfaces that enhance gas-fracture connectivity. Compared to slickwater, SC-CO_2 achieves 28–32% lower breakdown pressures (He et al., 2019; Jia et al., 2018) and penetrates micrometer-scale pores, creating fracture networks with greater branching density, 39% higher curvature, and permeability enhancements approaching 10^4–10^5 magnitudes. Unlike slickwater's bifurcated fractures, SC-CO_2 generates tri-wing geometries at high perforation angles. Its low viscosity and high diffusivity initiate early-stage micro-fractures that activate natural fracture networks, forming bedding-parallel fractures and complex interconnected systems (Ha et al., 2018; Shen et al., 2022).

Geological discontinuities in unconventional reservoirs significantly govern hydraulic fracturing outcomes. Fracture propagation fundamentally diverges in naturally fractured versus intact formations, as pre-existing fractures alter hydraulic fracture trajectories, promoting coalescence events that generate complex networks (Dehghan et al., 2015; Liu et al., 2023; Tan et al., 2023; Wang & Zhang, 2022; Wang et al., 2022; Xiong & Ma, 2022). While natural fracture activation enhances productivity, the resultant fracture merging complicates pressure analysis and growth prediction (Taleghani et al., 2016). Four primary interaction modes exist: net-crossing, net-opening, crossing-opening, and opening-crossing (Ru et al., 2020), with behavior governed by approach angles and natural fracture lengths. Under constant stress differentials, acute approach angles preferentially activate natural fractures (Guo et al., 2015b; Song et al., 2020), whereas extensive natural fracture networks direct propagation toward dominant orientations, enlarging connected fracture areas (Cao et al., 2022). Consequently, understanding fracture interactions proves critical for effective stimulation in pre-fractured formations, necessitating incorporation of natural fracture properties in predictive models of propagation paths and intersection geometries (Ghaderi et al., 2018; Wang et al., 2021).

Microseismic monitoring serves as a critical optimization tool for oilfield development, completion design, and stimulation strategies. Geomechanical analysis indicates microseismic hypocenters typically cluster near hydraulic fractures, enabling precise determination of fracture dimensions and propagation dynamics (Warpinski et al., 2013). Current methodologies characterize fractures through microseismic event localization (Yuan & Zhang, 2018), with both fluid pressurization and stress redistribution mechanisms generating diagnostic seismicity that elucidates fracture network characteristics (Maxwell et al., 2015). Microseismic data quantifies total fracture geometry (including height, length, azimuth) and maximum horizontal stress orientation, revealing interactions between completion operations and pre-existing fracture networks (Maxwell et al., 2009; Patterson et al., 2018). Field observations demonstrate direct correlation between injection pressure fluctuations and microseismic response (Maity & Ciezobka, 2019), while fluid volume governs maximum seismic magnitude. Peak-magnitude events typically postdate reservoir rupture or pump termination, corresponding to fracture shear mechanisms (Lu et al., 2018).

Multi-stage and multi-well fracturing technology has become integral for stimulating unconventional low-permeability reservoirs. Current research employs diverse methodologies to investigate fracture propagation dynamics during multi-well operations (Wang & Liu, 2022; Wang et al., 2023; Yang et al., 2023b; Yao et al., 2017; Zhang et al., 2021). Displacement discontinuity analyses reveal that fracture initiation and growth are governed by inter-well/cluster stress interference, resulting in asymmetric, dipped, and unbalanced propagation patterns (Yang et al., 2023a). Numerical simulations demonstrate stress-induced asymmetry where strong inter-well interference suppresses lateral growth of interior fractures (Chen et al., 2018). Hydraulic fractures accelerate when propagating toward tensile zones at adjacent fracture tips, while exhibiting distance-dependent inhibition on neighboring fractures (Ran et al., 2023). Laboratory studies confirm escalating fracture complexity with additional stages: inter-well fluid communication (frac-hit) reactivates existing fractures through fluid transfer, re-elevating induced stresses, though dimensional constraints limit most experiments to small scales (Guo et al., 2015a, 2023). Consequently, developing large-scale engineering models capable of accurately characterizing formation properties is imperative for elucidating fracture propagation mechanisms in multi-well low-permeability systems.

The remaining parts of this chapter are as follows: Sect. 8.2 introduces the hydrofracturing combined finite element-discrete element method based on THM coupling, discrete fracture network model (DFN), the fracturing numerical method for microseismic activity analysis, and related control equations (such as the governing equation for porous rock formations, seepage and fracture fluid flow and heat transfer), and microseismic activity analysis through torque tensor evaluation. Section 8.3 introduces the geometric and finite element models of multi-well fracturing considering THM coupling, as well as typical cases considering SC-CO_2 fracturing and natural fractures. Section 8.4 introduces the results and mechanism analysis of the fracture network morphology, fracture length, fracture volume, and induced microseismic events of the reservoir considering typical cases of THM coupling, SC-CO_2 fracturing, and natural fractures. Finally, Sect. 8.5 summarizes the main conclusions of this study.

8.2 Geomechanical Equations of Supercritical CO_2 Fracturing and Microseismic Analysis Considering Thermal-Hydro-Mechanical Coupling

8.2.1 Geomechanical Equations Considering Thermal-Hydro-Mechanical Coupling

The rock deformation, fluid seepage, fracture fluid flow, and heat transfer are coupled and considered in the model in this study to form thermal-hydro-mechanical coupling. The geomechanical equations are provided as below (Hantschel & Kauerauf, 2009; Wang et al., 2019).

$$\text{Rock deformation}: \quad \boldsymbol{L}^T(\boldsymbol{\sigma}^e - \alpha \boldsymbol{m} p_s) + \rho_b \boldsymbol{g} = 0 \tag{8.1}$$

$$\text{Fluid seepage}: \quad \text{div}\left[\frac{k}{\mu_l}(\nabla p_l - \rho_l g)\right] = \left(\frac{\phi}{K_l} + \frac{\alpha - \phi}{K_s}\right)\frac{\partial p_l}{\partial t} + \alpha \frac{\partial \varepsilon_v}{\partial t} \tag{8.2}$$

$$\text{Fracture fluid flow}: \quad \frac{\partial}{\partial x}\left[\frac{k^{fr}}{\mu_n}\left(\nabla p_n - \rho_{fn}\boldsymbol{g}\right)\right] = S^{fr}\frac{\mathrm{d}p_n}{\mathrm{d}t} + \alpha(\Delta \dot{e}_\varepsilon) \tag{8.3}$$

$$\text{Heat transfer}: \quad \text{div}\left[k_b \nabla T_f\right] = \rho_b c_b \frac{\partial T_f}{\partial t} + \rho_f c_f \boldsymbol{q}_f \nabla T_f \tag{8.4}$$

The heat transfer between the network nodes (fracturing fluid) and the formation nodes (rock matrix) are illustrated in Fig. 8.2.

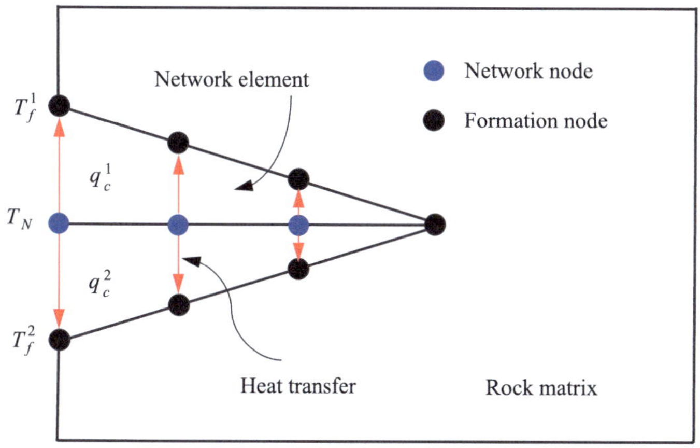

Fig. 8.2 Heat transfer via contact elements between the network and formation nodes

8.2.2 Microseismicity Analysis by the Evaluation of Moment Tensors

Through the analysis of the above coupling process, the evolved stress field can be obtained, and the microseismic behavior can be derived through the variation of stress field. The moment tensor $\boldsymbol{M}$ can be obtained by the Eigen decomposition via Eqs. (8.5) and (8.6). Using the moment tensor $\boldsymbol{M}$ at each moment, we can know the type of microseismic damage and contact-slip events in the whole region, the location, type and magnitude of microseismic events can be obtained.

$$\text{Eigen decomposition}: \quad \Delta\boldsymbol{\sigma}^e\boldsymbol{v} = \lambda\boldsymbol{v} \tag{8.5}$$

$$\text{Moment tensor}: \quad \boldsymbol{M} = \lambda\boldsymbol{v} \tag{8.6}$$

Some details of above governing equations are not the focus of this study, that can be found in previous studies (Wang et al., 2021), in which the related methods (e.g., numerical discretization, fracture criterion, and leak-off) are not described here. Especially, the thermal conductivity, specific heat, and incremental volume in heat transfer in this study are given in previous chapters. The involved symbols and physical meaning of parameters are summarised in Table 8.1. Table 8.2 shows the meanings and values of the physical parameters in these equations. The fluid and heat parameters for slick water and SC-CO_2 fracturing are listed in Table 8.3.

Table 8.1 Symbols and physical meaning of parameters

Symbol	Physical meaning	Symbol	Physical meaning
$\boldsymbol{L}$	Spatial differential operator	k_b	Thermal conductivity
$\boldsymbol{\sigma}^e$	Effective stress tensor	T_f	Fluid temperature
$\boldsymbol{m}$	Identity tensor	$\boldsymbol{q}_f$	Darcy fluid flux
$\boldsymbol{g}$	Gravity vector	λ	Eigenvalue
ε_v	Volumetric strain of the rock formation	$\boldsymbol{v}$	Eigenvector
t	Time	$\boldsymbol{M}$	Moment tensor
$\Delta\dot{e}_\varepsilon$	Aperture strain rate		

Table 8.2 Basic physical parameters

Parameter	Value
Temperature of rock stratum /°C	60
Depth of horizontal wells/m	3000
Fluid injection volume in one fracturing stage/m^3	50
Fluid injection duration in one fracturing stage/s	500
Fluid injection rate Q/(m^3/s)	0.1
Leak-off coefficient C_I/($m^3/s^{1/2}$)	0.1×10^{-15}
Leak-off coefficient C_{II}/($m^3/s^{1/2}$)	0.1×10^{-15}
Pore pressure p_s/MPa	30
Porosity ϕ	0.05
Gravity g/(m/s^2)	9.81
Horizontal minimum in-situ stress in x direction S_h/MPa	40
Horizontal maximum in-situ stress in y direction S_H/MPa	60

8.2.3 Discrete Fracture Network Model

In 2D DFN, a fracture is represented as a line segment connecting two points in the region:

$$H \equiv [P, Q] \tag{8.7}$$

where $P = (x_1, y_1)$, and $Q = (x_2, y_2)$ are the endpoints of H. We can also represent a fracture using four parameters, $w = (w_1, w_2, w_3, w_4)$, where (w_1, w_2) are the coordinates of the center and w_3 and w_4 are, respectively, the orientation and the length of the fracture (Seifollahi et al., 2014).

Table 8.3 Fluid and heat parameters for slick water and SC-CO_2 fracturing

Parameters	Value
Dynamic viscosity coefficient of the pore fluid μ_g/(Pa · s)	1.00×10^{-3}
Dynamic viscosity coefficient of slick water μ_n/(Pa · s)	1.67×10^{-3}
Dynamic viscosity coefficient of SC-CO_2 μ_n/(Pa · s)	1.67×10^{-5}ara>
Density of the pore fluid ρ_g/(kg/m^3)	1×10^3
Density of slick water ρ_f/(kg/m^3)	1×10^3
Density of SC-CO_2 ρ_f/(kg/m^3)	0.66×10^3
Bulk modulus of the pore fluid K_g/MPa	2050
Bulk modulus of slick water K_f^{fr}/MPa	2000
Bulk modulus of SC-CO_2 K_f^{fr}/MPa	58
Grain thermal conductivity k_s/[W/(m·K)]	2.9
Bulk specific heat c_b/[J/(kg·°C)]	440.5
Fluid specific heat c_f/[J/(kg·°C)]	600
Linear thermal coefficient of propagation α_t/K^{-1}	2×10^{-5}
Contact thermal conductivity α_c/[W/(m·°C)]	58.8

DFN model gives more practical consideration to the fracture occurrence by fully discretizing the matrix and fracture system in the modelling process, and can more accurately describe the flow law of reservoir under heterogeneous conditions. As shown in Fig. 8.3, the fractures are discretized by finite element method, and the matrix is discretized by triangular elements with the fracture elements serving as the grid constraint (Karimi-Fard & Firoozabadi, 2001).

In this method, the fractures are reduced and homogenized. The discrete DFN model can be applied to any complex fracture structure theoretically by

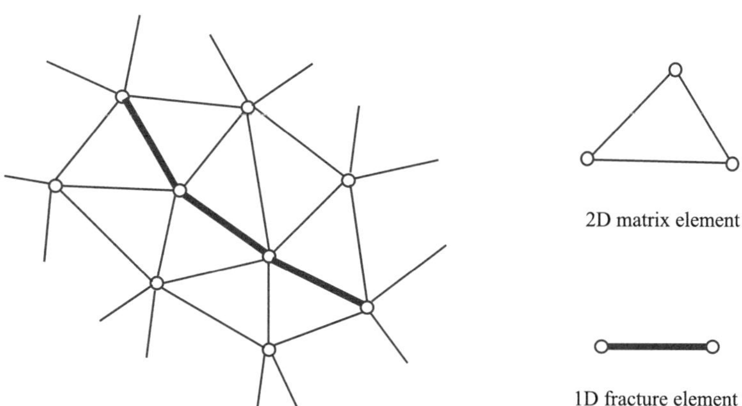

Fig. 8.3 Discretization method of discrete fractured media

$$\int_{\Omega} f(x, y)\mathrm{d}\Omega = \int_{\Omega_m} f(x, y)\mathrm{d}\Omega_m + \int_{\Omega_f} f(x, y)\mathrm{d}\Omega_f$$
$$= \int_{\Omega_m} f(x, y)\mathrm{d}\Omega_m + e \times \int_{\overline{\Omega}_f} f(x, y)\mathrm{d}\overline{\Omega}_f \tag{8.8}$$

where $f(x, y)$ represents the flow equations of the rock matrix and the fracture media; Ω represents the global domain; Ω_m represents the rock matrix; Ω_f represents the fracture; $\overline{\Omega}_f$ represents the fracture part of the domain as a 1D; and the fracture width is represented by e, which is the coefficients before one dimensional integration.

8.3 Numerical Models of Supercritical CO_2 Fracturing in Fractured Reservoir

8.3.1 Geometrical and Finite Element Models

This study employs a 600 m × 400 m two-dimensional discrete fracture network (DFN) model (Fig. 8.4) containing three horizontal wells with 100 m inter-well spacing. Each well features five perforation clusters spaced at 100 m intervals, with initial fracture lengths of 2 m. Sequential fracturing progresses bottom-to-top across wells and left-to-right within perforation clusters. The naturally fractured reservoir domain exhibits localized pre-existing fractures (500 m × 300 m) surrounding clusters rather than global distribution. To optimize computational efficiency and solution reliability, adaptive finite element mesh refinement dynamically focuses on fracture propagation fronts, ensuring accurate stress solutions at fracture tips compared to conventional methods. Thermally, fracturing fluids within wells and fracture networks maintain distinct temperatures from the rock matrix. Table 8.4 details stage durations and total operational times for four uniformly executed simulation cases, with each well's five-stage sequence completing within 2500 s per well.

Computational accuracy is enhanced through localized mesh refinement around perforation clusters, implementing finer initial finite elements within a 500 m × 300 m domain (Fig. 8.5). Adaptive mesh refinement during fracture propagation optimizes modeling efficiency while ensuring reliable stress solutions at fracture tips and along propagation paths outperforming conventional approaches. Table 8.5 details refinement and coarsening parameters for multi-well fracturing simulations within this finite element framework.

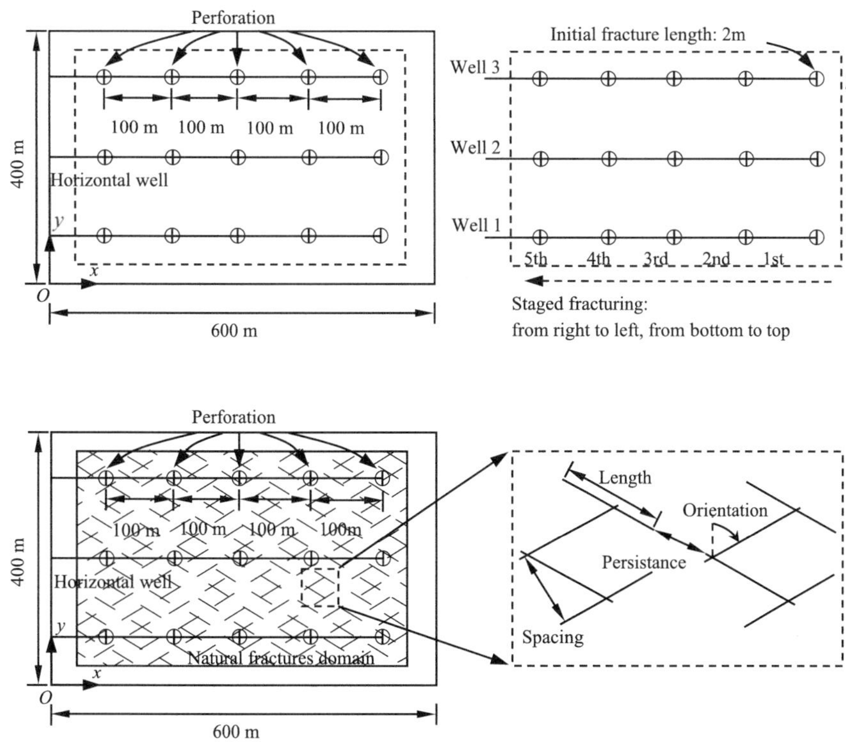

Fig. 8.4 Geometrical models of multi-well fracturing in unfractured and naturally fractured reservoirs

8.3.2 *Cases Study for Typical Fracturing Fluids: Slick Water and Supercritical CO_2*

Discrete Fracture Networks (DFNs) model pre-existing fractures to isolate their influence on hydraulic fracture propagation. Each fracture set is characterized by orientation (degrees clockwise from vertical), spacing (perpendicular distance between adjacent fractures), length (tip-to-tip distance), and persistence (longitudinal separation between fracture termini), which parameters governing fracture spatial distribution (Fig. 8.4). Table 8.6 details the constitutive properties for two such fracture sets.

8.4 Results and Discussion

In this section, the above-mentioned 16 numerical cases for comparative analysis are computed using the proposed models, and ELFEN TGR (Rockfield Software Ltd., 2016) is utilized for four cases on a desktop computer with an Intel® Core™

Table 8.4 Duration and total time of fracturing stages

Stage		Duration/s	Total time/s
Initial balance		2.00	2.00
1st well	1st stage	500	502 (0.13944 h)
	2nd stage	500	1002 (0.27833 h)
	3rd stage	500	1502 (0.41722 h)
	4th stage	500	2002 (0.55611 h)
	5th stage	500	2502 (0.695 h)
2nd well	1st stage	500	3002 (0.83389 h)
	2nd stage	500	3502 (0.97278 h)
	3rd stage	500	4002 (1.1117 h)
	4th stage	500	4502 (1.25056 h)
	5th stage	500	5002 (1.38944 h)
3rd well	1st stage	500	5502 (1.52833 h)
	2nd stage	500	6002 (1.66722 h)
	3rd stage	500	6502 (1.80611 h)
	4th stage	500	7002 (1.945 h)
	5th stage	500	7502 (2.08389 h)

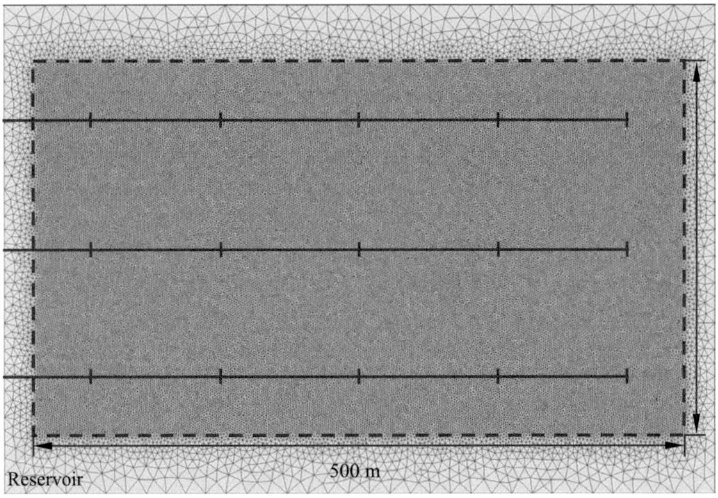

Fig. 8.5 Finite element model of multi-well SC-CO_2 fracturing in naturally fractured reservoirs

3.40 GHz CPU. The following section discusses the fracture propagation and intersection behaviors in unfractured and naturally fractured reservoirs with different fracturing fluids and analyses the results of quantitative fracture length, volume, and microseismic damage and contact slip events.

Table 8.5 Computational parameters for mesh refinement and coarsening

Parameters	Value
Initial intensive region in x direction/m	$50m \leq x \leq 550m$
Initial intensive region in y direction/m	$50m \leq y \leq 350m$
Small detail size	2
Fracture mesh size factor	0.5
Mesh density factor	1
Mesh density	0.125
Bubble size	3
Coarsening frequency	10
Coarsening density factor	1
Coarsening density	0.5
Coarsening threshold factor	0.9
Coarsening threshold	0.45
Non-coarsening zone factor	30
Non-coarsening zone	15
Max coarsening zone factor	15
Max coarsening zone	7.5

Table 8.6 Pre-existing fracture sets of the naturally fractured model

State	DFN	
	Set 1	Set 2
Orientation/(°)	60	120
Spacing/m	15	15
Length/m	15	15
Persistence/m	15	15

8.4.1 Intersections and Connections of Fracturing Fracture Networks

In this section, numerical cases of slick water fracturing (Case I and Case III) and SC-CO_2 fracturing (Case II and Case IV) in unfractured and naturally fractured reservoirs are computed. Table 8.7 lists the initial temperatures of the fracturing fluid in different cases. When the horizontal well depth in the deep dense oil reservoir is 3000 m, the rock temperature is set at 60 °C. It should be noted that if the cluster spacing is significantly reduced (such as within the range of 20–50 m), the stress shadow effect will significantly affect the propagation behaviour of SC-CO_2 fracturing fractures. The cluster spacing holding 100 m weakens the interaction between fractures, which allows us to focus on heat transfer and natural fractures.

Table 8.7 Fracturing fluids and initial temperatures of different cases

Cases	I	II	III	IV
Fracturing fluid	Slick water	SC-CO_2	Slick water	SC-CO_2
Temperature of rock stratum/°C	60	60	60	60
Temperature of fracturing fluid/°C	20	35	20	35

To analyze the final fracture network morphology and first principal stress during multi-well fracturing, multi-stage fracturing using slick water and SC-CO_2 are computed and analysed. Figure 8.6 shows the morphology of fracturing fractures in four different cases and provides the first principal stress. Figures 8.6a and b show the final fracture fracture network morphology of unfractured reservoirs fractured using slick water and SC-CO_2, with typical single fracturing fracture occurring. Because the cluster spacing is relatively large (i.e., 100 m), the stress shadow effect between clusters is relatively weak. Therefore, the propagation of fracturing fractures is almost parallel, which is the same as multi-stage hydrofracturing of single horizontal well. As shown in Fig. 8.6a, for slick water fracturing in unfractured reservoir, due to the influence of well 1 on the in-situ environment of the reservoir, fractures 1 and 4 of well 2 connected with well1. Due to the influence of well 1 and well 2 on the in-situ environment of the reservoir, fractures 1, 2, and 3 of well 3 are connected with well 2. As shown in Fig. 8.6b, for SC-CO_2 fracturing in unfractured reservoir, due to the influence of well 1 on the in-situ environment of the reservoir, fractures 1 and 4 of well 2 connected with well 1. Due to the influence of well 1 and well 2 on the in-situ environment of the reservoir, fractures 2 and 3 of well 3 are connected to well 2. Figures 8.6c and d show the final fracturing fracture network of a naturally fractured reservoir using slick water and SC-CO_2 fracturing. It can be seen that fracturing fractures propagate and intersect with pre-existing natural fractures, forming a complex fracture network. Similar to unfractured reservoirs, the stress shadow effect between clusters in naturally fractured reservoirs is still weak. The comparison results of the fracture network using slick water and SC-CO_2 fracturing show that there are some differences in the propagation of the corresponding clusters. The fracturing fractures using SC-CO_2 fracturing are significantly longer than those using slick water. As shown in Fig. 8.6c, during the fracturing process using slick water, due to the influence of well 1 on the in-situ environment of the reservoir, fracture 5 of well 2 connected with well 1. During the fracturing process using SC-CO_2, due to the influence of well 1 on the in-situ environment of the reservoir, fracture 5 of well 1 connected with well 2. Due to the influence of well 1 and well 2 on the in-situ environment of the reservoir, fractures 3 and 4 of well 3 were connected to well 2, as shown in Fig. 8.6d. Therefore, the properties of fracturing fluid are important factors affecting fracturing performance.

For SC-CO_2 fracturing in naturally fractured reservoirs (Case IV), the evolution of shear stress σ_{xy} near fracture 2 of well 1 is shown in Fig. 8.7. From the stress results of the first stage fracturing of well 1, as shown in Fig. 8.7a, positive shear stress is caused on the left side of the upper fracture tip, while negative shear stress is

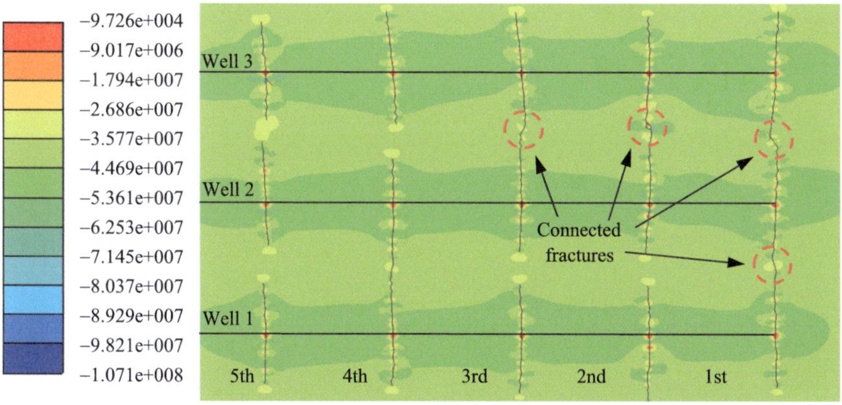

(a) Slick water fracturing in unfractured reservoir

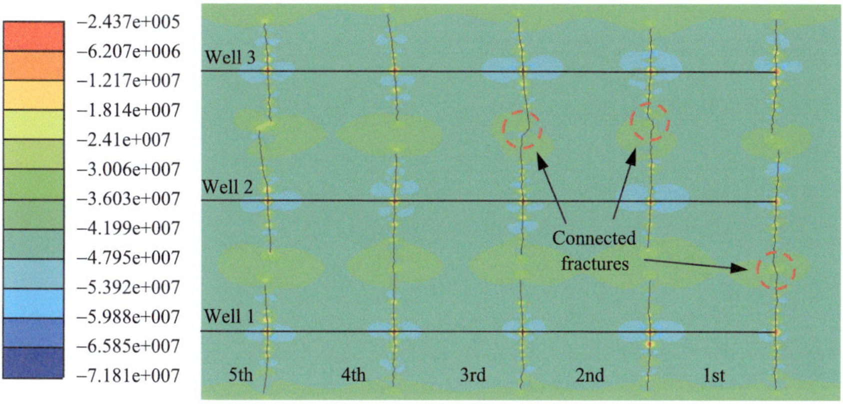

(b) SC-CO_2fracturing in unfractured reservoir

Fig. 8.6 Results of stress (first principal stress, Pa) and fracture morphology of unfractured and naturally fractured model

caused on the right side; the left side of the lower fracture tip causes negative shear stress, while the right side of the lower fracture tip causes positive values. In addition, the stress around fracture 1 is not symmetrical, which is caused by the pre-existing natural fractures during the multi-well fracturing process. Subsequently, from the stress results of the second stage fracturing of well 1, as shown in Fig. 8.7b, fracture 2 begins to propagate. The left and right sides of the upper and lower fracture tips causes positive and negative shear stresses, respectively, as shown in the patterns of fracture 1. These stress distributions are typical patterns of the shear stress field around the fracture. As shown in Fig. 8.7c, the area of the shear stress field variations caused by the propagation of fracture 1 partially covers the variation area induced by fracture 2. Therefore, the shear stress field caused by the two fractures leads to superposition and reduction. The shear stress field on the right side of fracture 2 is

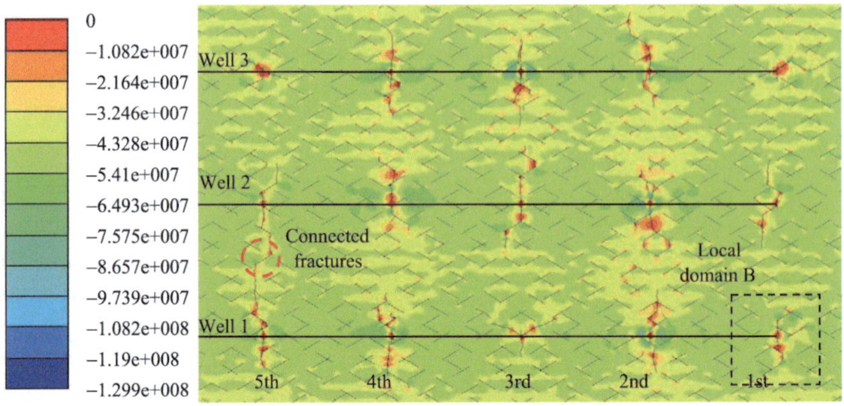

(c) Slick water fracturing in naturally fractured reservoir

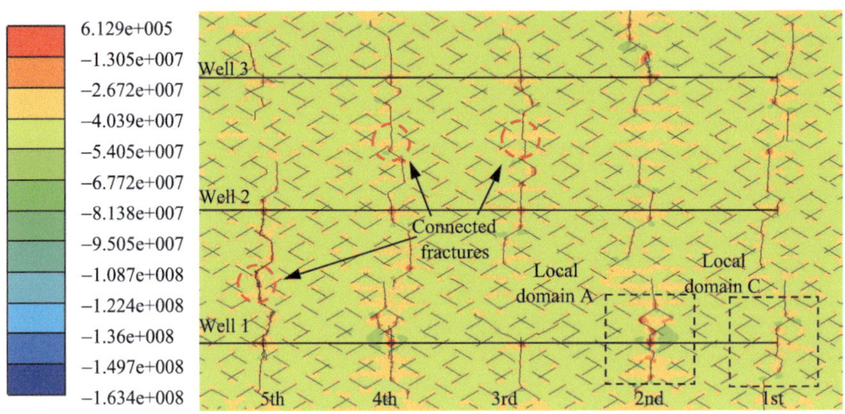

(d) SC-CO_2fracturing in naturally fractured reservoir

Fig. 8.6 (continued)

relatively weakened. Under the joint effect of natural fractures and shear stress field, fracture 2 begins to propagate and deflect towards the left side of the larger stress area, and forms the final propagation morphology, as shown in Fig. 8.7d. From these results, it can be seen that during the fracturing process, the initial in-situ stress is disturbed by natural fractures and fracture propagation, while stress concentration increases at the fracture tip. The stress concentration areas will result in fracture deflection through stress superposition and reduction of the interaction, i.e., the stress shadow effect. It should be noted that if the cluster spacing is significantly reduced (such as within the range of 20–50 m), the stress shadow effect will significantly affect the propagation behaviour of SC-CO_2 fracturing fractures. The 100-m cluster spacing weakens the interaction between fractures, allowing us to focus on issues related to heat transfer and natural fractures.

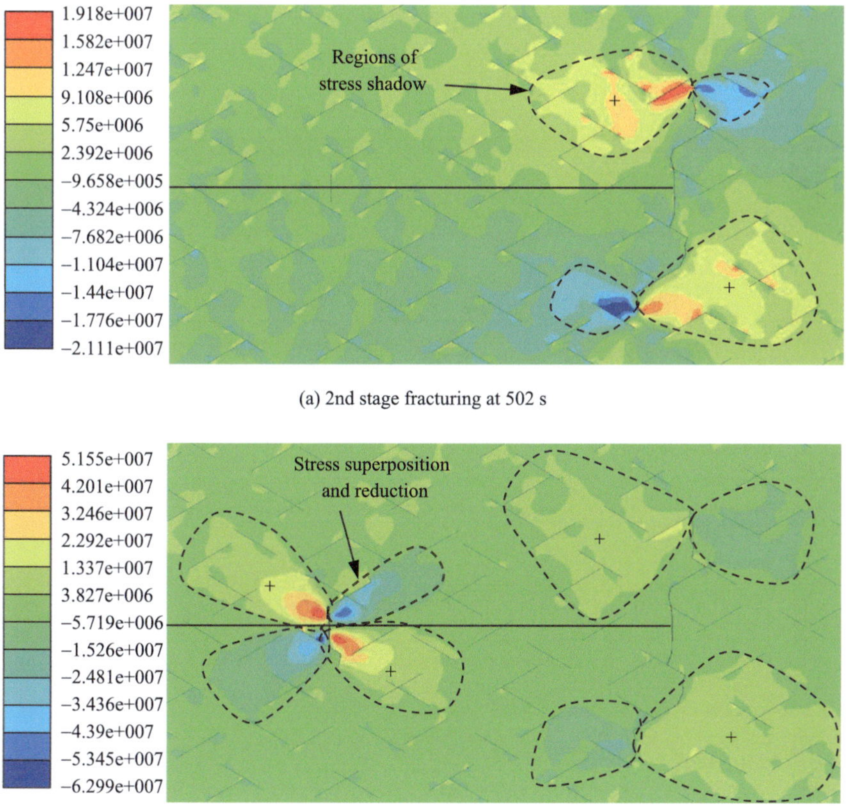

(a) 2nd stage fracturing at 502 s

(b) 2nd stage fracturing at 652 s

Fig. 8.7 Dynamic evolution of shear stress and fracture morphology of fractures 2 of well 1 (local domain A)

To observe the local changes more carefully, Fig. 8.8 shows the temperature distribution and thermal gradient of heat conduction around the fracture network of selected local domain B and C at 502 s. Figures 8.8a and b show the temperature distribution around the fracture network of naturally fractured reservoir using slick water and SC-CO_2 fracturing. It can be observed that fracturing fractures propagate and connect to pre-existing natural fractures, forming a complex network of fractures, with high-temperature domains near the bilateral regions of the fracture. During the propagation of fractures, temperature radiates outward from the initiation point of the fracture (injection point of fracturing fluid). The temperature at the initiation point is close to the temperature of the fracturing fluid, and the temperature in the high-temperature zone is close to the temperature of the rock formation. There is a heat transfer zone between the two mentioned above. Heat transfer occurs along fractures and pre-existing natural fractures, and the temperature range of SC-CO_2 fracturing (due to temperature difference caused by heat transfer) is smaller than

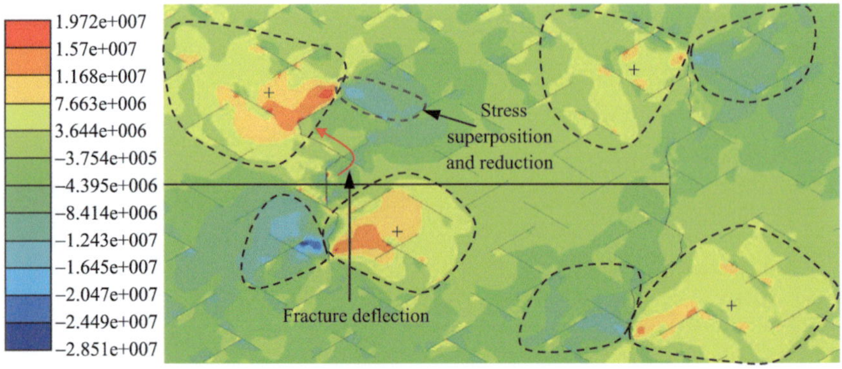

(c) 2nd stage fracturing at 802 s

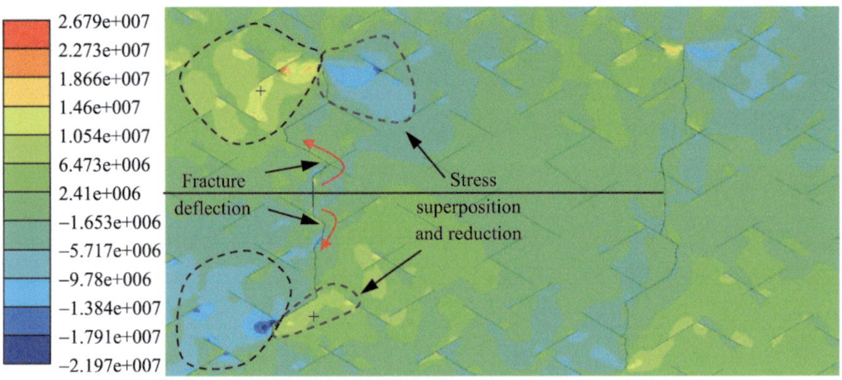

(d) 2nd stage fracturing at 1002 s

Fig. 8.7 (continued)

that of slick water fracturing because the initial temperature difference of SC-CO_2 fracturing (SC-CO_2 at 35 °C and rock formation at 60 °C) is smaller than that of slick water fracturing (slick water at 20 °C and formation at 60 °C). Figures 8.8c and d show the thermal gradient of thermal conduction around the fracture network of naturally fractured reservoir with slick water and SC-CO_2. The thermal gradient of heat conduction varies significantly near the initial perforation area. The thermal gradient of slick water fracturing is significantly higher than that of SC-CO_2 fracturing. The comparison results of the fracture network using slick water and SC-CO_2 fracturing in Fig. 8.8 show some differences in the fracture propagation of the corresponding clusters. Therefore, the properties of fracturing fluid are important factors affecting fracturing performance.

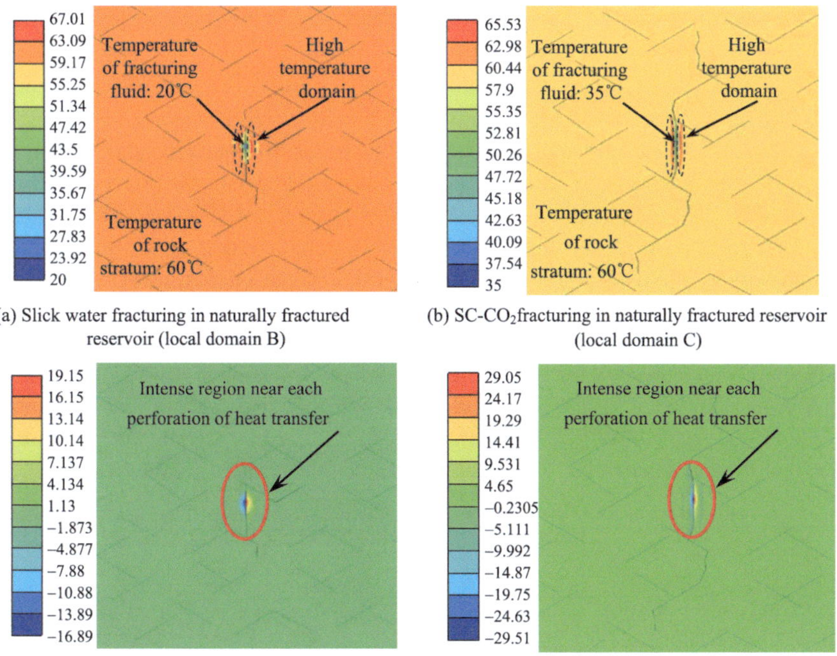

(a) Slick water fracturing in naturally fractured reservoir (local domain B)

(b) SC-CO_2fracturing in naturally fractured reservoir (local domain C)

(c) Slick water fracturing in naturally fractured reservoir (local domain B)

(d) SC-CO_2fracturing in naturally fractured reservoir (local domain C)

Fig. 8.8 Distribution of thermal gradient in the x direction around the fracturing fracture network in 1st stage fracturing of 1st well at time 502 s

8.4.2 Quantitative Variation of Fracture Networks, Fluid Rate, and Pore Pressure

To investigate the influences of different fracturing fluids on multi-well fracturing in unfractured and naturally fractured reservoirs, the quantitative results of the fracture network may be derived and discussed in detail as follows.

Figure 8.9 shows the evolution of the total fracture length and volume of the fracturing fractures. Fracturing fractures in naturally fractured reservoirs do not include the connected pre-existing fractures. It can be seen that the increasing trend of fracture length and volume is similar in the four cases. During the overall fracturing process, the fracture length monotonically increases, while the fracture volume significantly decreases during the flowback stage. However, for both slick water fracturing and SC-CO_2 fracturing, the fracture length of naturally fractured reservoirs is greater than the value of unfractured reservoirs, while the fracture volume is not significantly different. The reason may be that a developed fracturing fracture network has been formed in the natural fracture reservoir, which is beneficial for increasing the length of fractures. After the completion of the fracturing stage, the closure of the fracturing fractures and pre-existing fractures leads to a sharp decrease of the fracture volume. In addition, for natural fractured reservoirs, the fracture length using SC-CO_2 fracturing is significantly longer than that using slick water. This is because natural fractures increase the connectivity and complexity of the original reservoir, and SC-CO_2, as a fracturing fluid, makes it easy to connect macroscopic and microcosmic fractures, forming a complex fracture network. The increase in fracture length is caused by the continuous injection of fracturing fluid, forming a network of fractures that provides seepage channels for natural gas extraction. It should be noted that the delivery of proppant in the fracture network and the improvement of proppant quality will support the fracture and result in a larger fracture volume.

Figure 8.10 shows the evolution of the fluid rate of multistage hydrofracturing in unfractured and naturally fractured reservoir. For slick water and SC-CO_2 fracturing, there is no significant difference in fluid velocity between the two fracturing fluids.

Due to the leak-off of the fracturing fluid, in all four cases, the fluid flow rate in the fracture was lower than the injection rate of the fracturing fluid (0.1 m^3/s). In

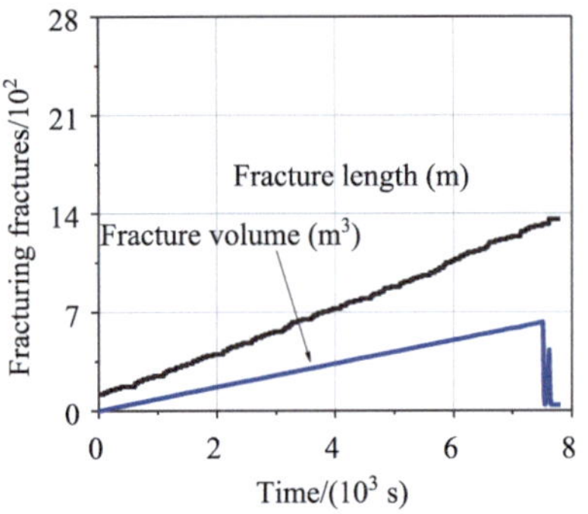

(a) Slick water fracturing in unfractured reservoir

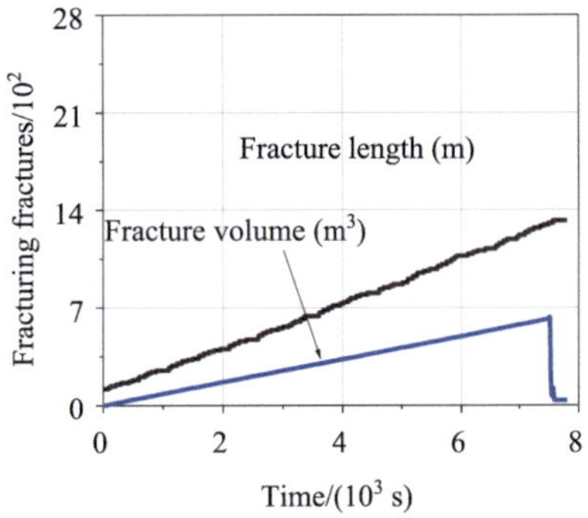

(b) SC-CO_2fracturing in unfractured reservoir

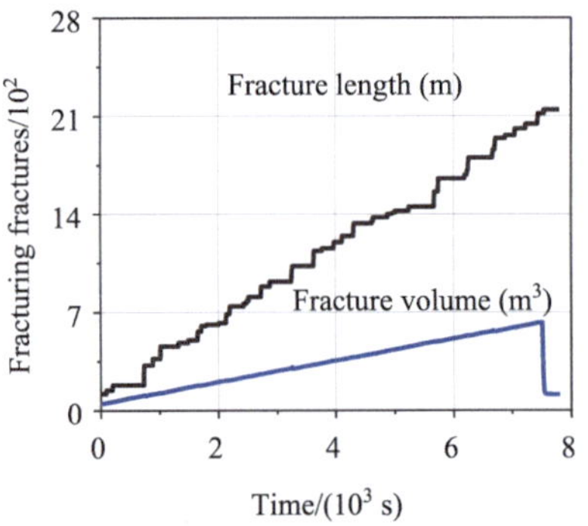

(c) Slick water fracturing in naturally fractured reservoir

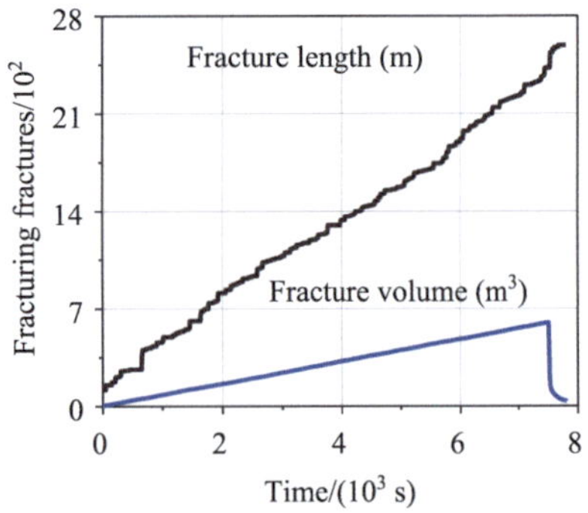

(d) SC-CO_2fracturing in naturally fractured reservoir

Fig. 8.9 Evolution of total fracture length and volume of the fracturing fractures

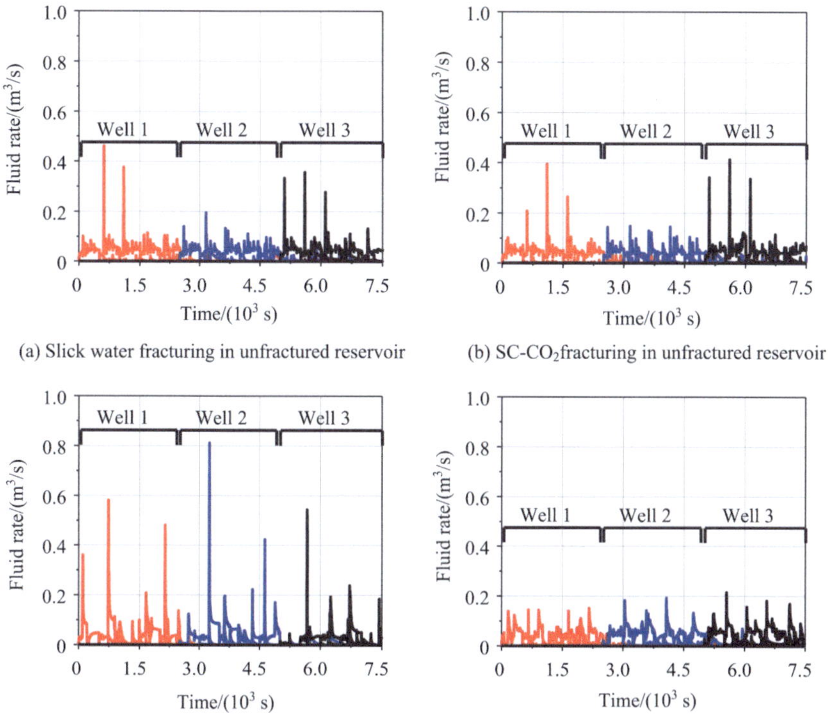

(a) Slick water fracturing in unfractured reservoir

(b) SC-CO_2fracturing in unfractured reservoir

(c) Slick water fracturing in naturally fractured reservoir

(d) SC-CO_2fracturing in naturally fractured reservoir

Fig. 8.10 Evolution of fluid rate of multistage hydrofracturing in unfractured and naturally fractured reservoir

addition, the injection rate of fracturing fluid is constant, therefore, the injection rate at each stage is almost constant and stable within a certain range. However, during the transformation of the fracturing stage, it can be observed that there is an abrupt increase at the beginning of each fracturing stage. After switching the fracturing stage, a newly emerged cluster has accumulated fracturing fluid, causing the fluid rate to continuously increase and disrupting the stable state until the fluid pressure in the cluster drove the initiation of the fracture. This is the reason for the abrupt increase phenomenon in the fluid rate observed at the beginning of each fracturing stage.

Figure 8.11 shows the evolution of pore pressure of multistage hydrofracturing in unfractured and naturally fractured reservoir. For unfractured reservoirs, there is no significant difference in pore pressure between slick water and SC-CO_2 fracturing. For naturally fractured reservoirs, the pore pressure of SC-CO_2 fracturing is slightly higher than that of slick water fracturing. In addition, due to the presence of pre-existing fractures in the reservoir, fluids are prone to diffusion in naturally fractured reservoirs, resulting in smaller pore pressure than unfractured reservoirs. Similar to

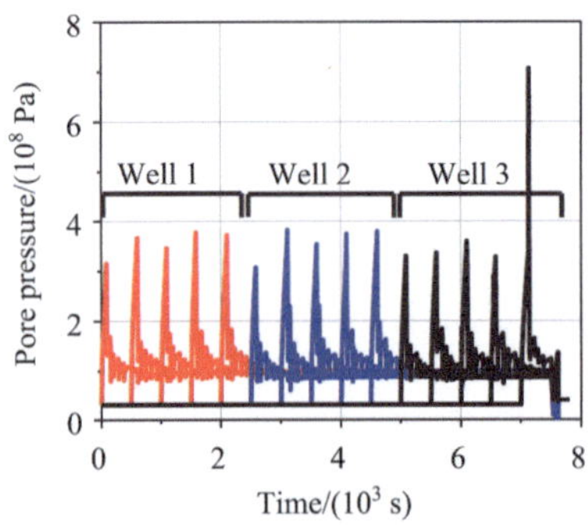

(a) Slick water fracturing in unfractured reservoir

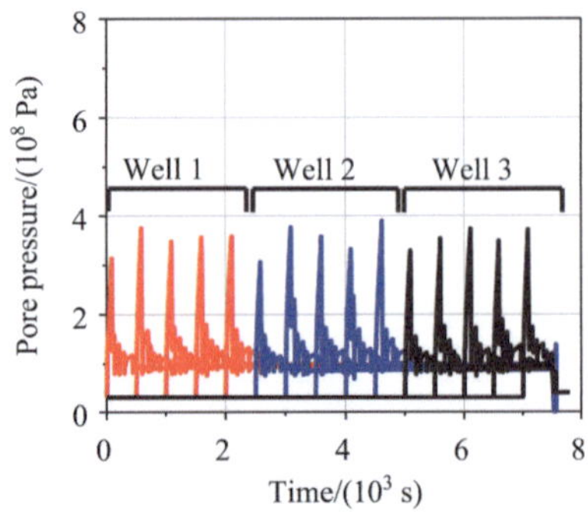

(b) SC-CO_2fracturing in unfractured reservoir

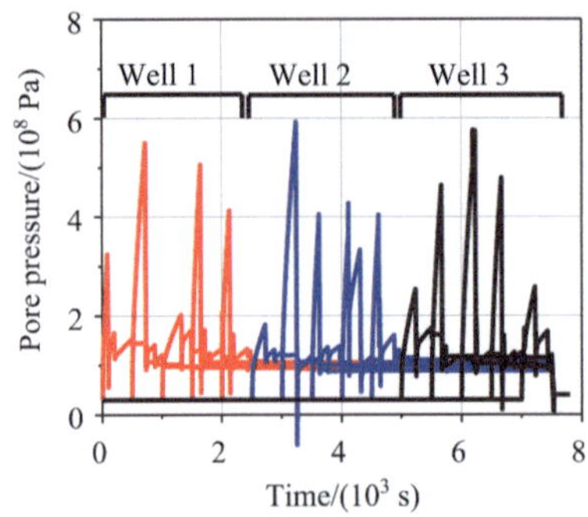

(c) Slick water fracturing in naturally fractured reservoir

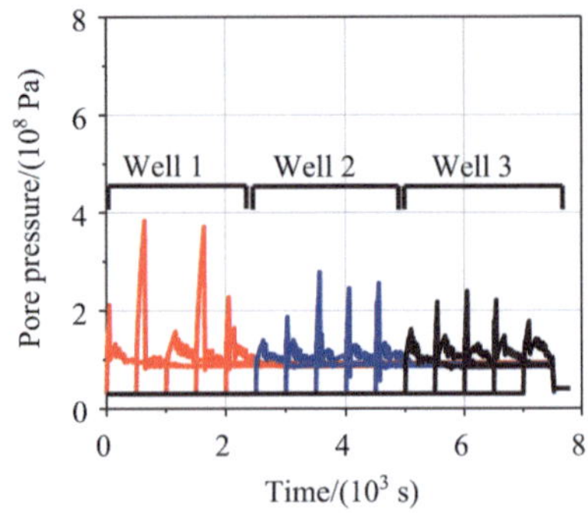

(d) SC-CO_2fracturing in naturally fractured reservoir

Fig. 8.11 Evolution of pore pressure of multistage hydrofracturing in unfractured and naturally fractured reservoir

the fluid rate, as the fracturing stage switches, the pore pressure suddenly increases at the beginning of each fracturing stage due to the storage of the fracturing fluid, showing the same trend as the sudden increase of the fluid rate at each stage, as shown in Fig. 8.10.

8.4.3 Microseismic Damage and Contact-Slip Events

By utilizing computed results, such as those of the effective stress, the moment tensors can be evaluated and microseismicity can be detected. To conduct comparative research, important results were obtained on the distribution, maximum magnitudes, and accumulated magnitudes of damage and contact slip events in slick water and SC-CO_2 fracturing for the four cases. The microseismic damage events caused by tensile failure are most likely to occur on the surface of fracturing fractures, mainly controlled by the evolution of the first principal stress, which mainly induces fracture. Microseismic contact slip events caused by shear slip failure are most likely to occur on pre-existing natural fracture surfaces, as these surfaces are prone to slide and tear due to the combined effects of injected fracturing fluid and in-situ stress. The focal spheres of the global distribution of final damage events, contact slip events, and first principal stress (Pa) through microseismic analysis are shown in Fig. 8.12. For unfractured reservoirs, the damage and contact slip events of slick water and SC-CO_2 fracturing are distributed along a single fracturing fracture at each fracturing stage, as shown in Figs. 8.12a–d. However, in slick water and SC-CO_2 fracturing, the number of contact slip events is significantly less than damage events, indicating that hydrofracturing has fewer shear fractures than tensile fractures. These results are consistent with the statistical results obtained from multi-stage hydrofracturing of a single well. To demonstrate the occurrence of local microscopic phenomena, we paid special attention to some local domains (D, E, F and G). The distributions of final damage events, contact slip events, and first principal stress in these selected representative domains are shown in Fig. 8.13. Figure 8.13a shows the damage events in the fracture intersection area of slick water fracturing in unfractured reservoir, which clearly occur with the formation of hydraulic fractures. Figure 8.13b shows the contact slip events in the fracture intersection area of SC-CO_2 fracturing in unfractured reservoir. For naturally fractured reservoirs, as shown in Fig. 8.12e–h, the damage events of slick water and SC-CO_2 fracturing are accompanied by new fracturing fractures and occur between pre-existing natural fractures. At each fracturing stage, there are more damage events in slick water fracturing than in SC-CO_2 fracturing. The contact slip event is distributed along the connected natural fractures and occurs at the intersection of far-field natural fractures. Figure 8.13c shows the damage events in the fracture intersection area of slick water fracturing in natural fractured reservoirs, Fig. 8.13d shows the contact slip events in the fracture intersection area of SC-CO_2 fracturing in natural fractured reservoirs. It can be clearly seen that there are more damage events than contact slip events, and the contact slip events distributed along SC-CO_2 fracturing fractures are significantly less than those distributed along natural fractures.

The evolution of the maximum and accumulated magnitudes of damage and contact slip events are shown in Fig. 8.14. It can be observed that in the process of multi-well fracturing, damage and contact slip events occur simultaneously in reservoir, with significantly more damage events than contact slip events. The maximum and accumulated magnitudes of damage events are significantly larger than those of the contact slip events. The gradual propagation of fracturing fractures will generate many induced microseismic events with large magnitude, and these damage and contact slip events cause up-and-down undulation in the magnitude of microseismic events. The damage events occur steadily from hydrofracturing, and the occurrence of more contact slip events in the initial perforation area increases the number and magnitude of contact slip events. As shown in Fig. 8.14a–d, for unfractured reservoirs, the up-and-down undulation caused by SC-CO_2 fracturing is weaker than that caused by slick water fracturing. As shown in Fig. 8.14e–h, for naturally fractured reservoirs, the up-and-down undulation caused by SC-CO_2 fracturing is similar to that caused by slick water fracturing due to the influence of natural fractures. According

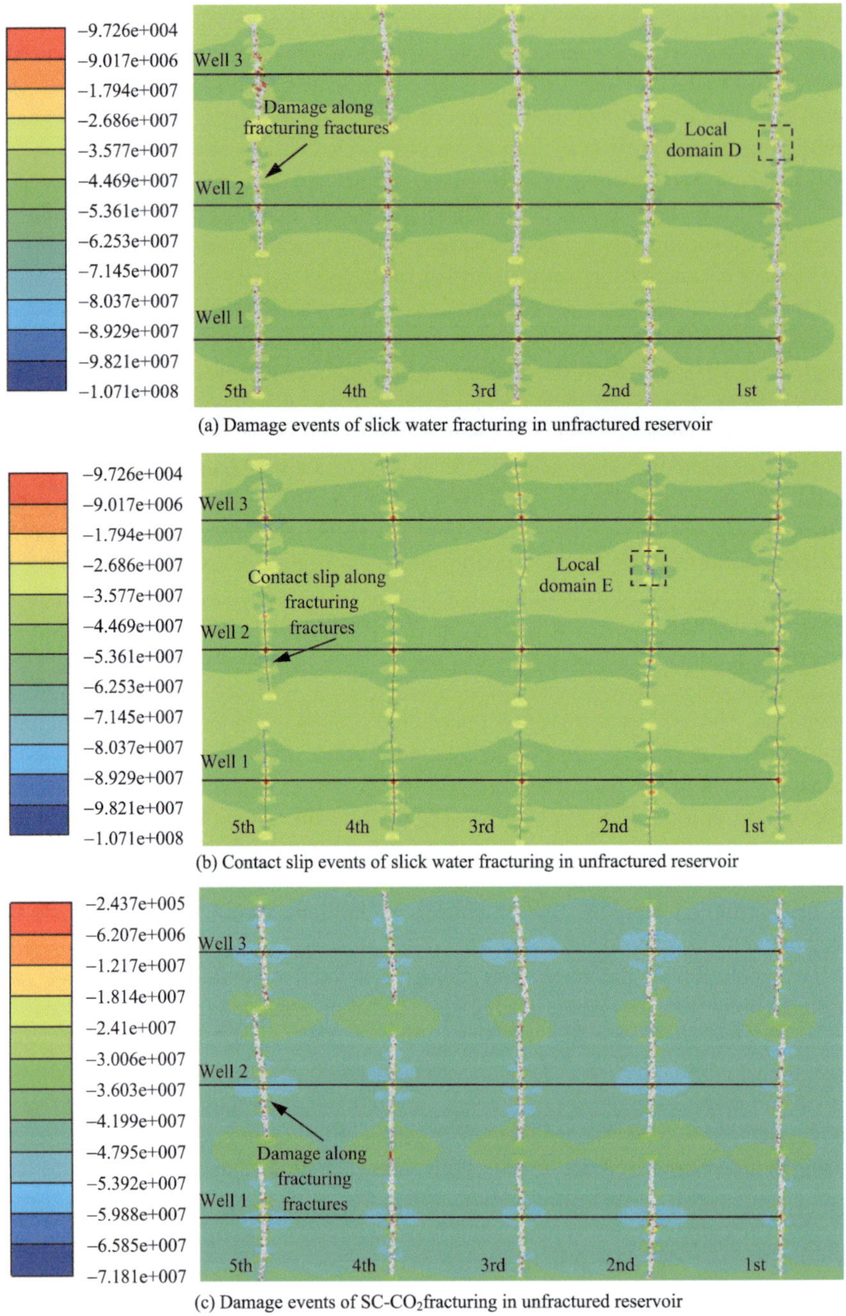

Fig. 8.12 Distribution of the final damage events, contact slip events, and the first principle stress (Pa) by microseismic analyses in the global domain

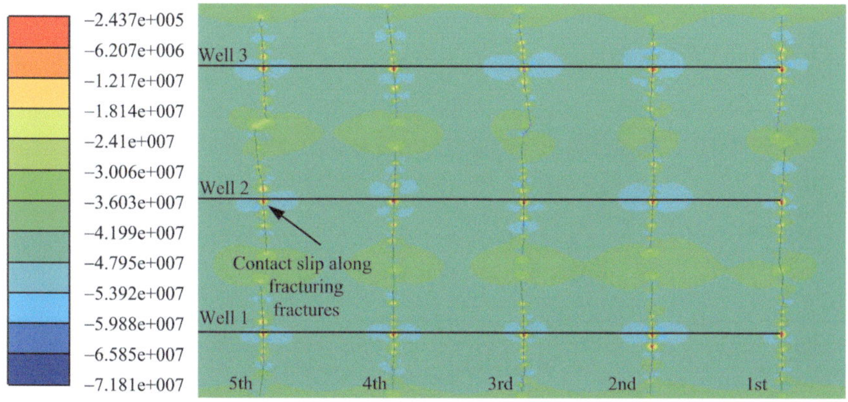

(d) Contact slip events of SC-CO_2fracturing in unfractured reservoir

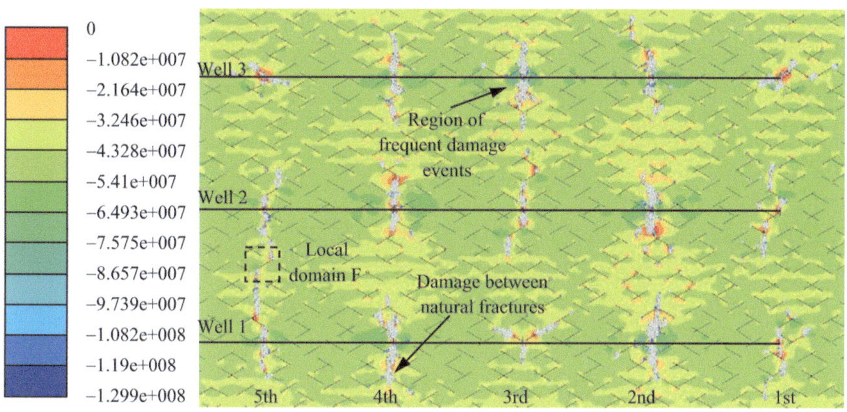

(e) Damage events of slick water fracturing in naturally fractured reservoir

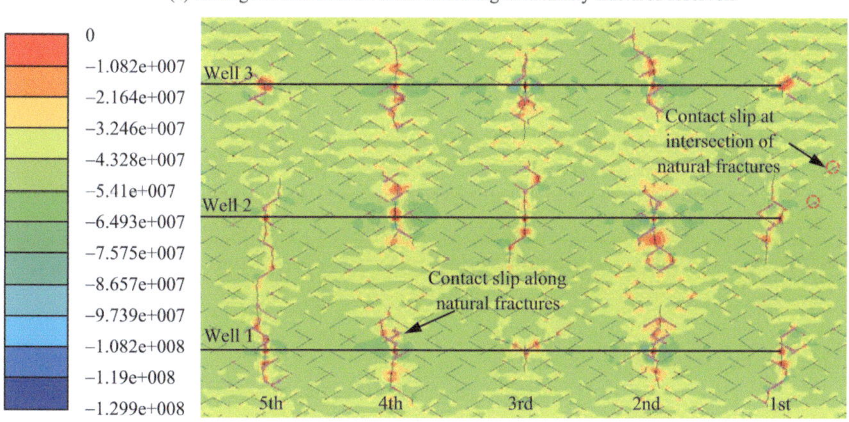

(f) Contact slip events of slick water fracturing in naturally fractured reservoir

Fig. 8.12 (continued)

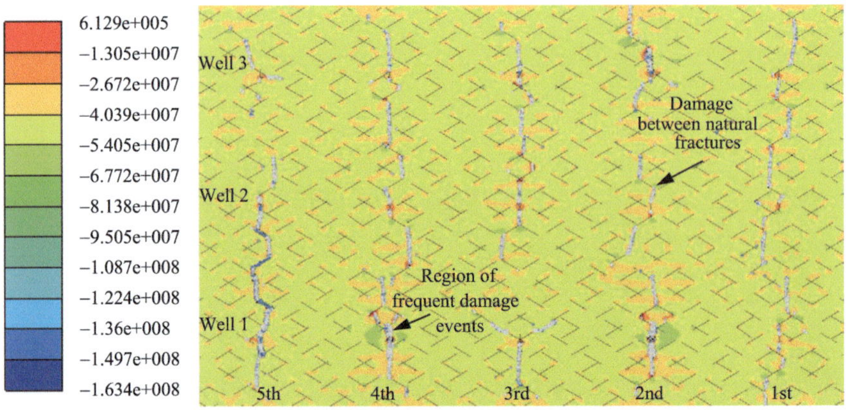

(g) Damage events of SC-CO_2 fracturing in naturally fractured reservoir

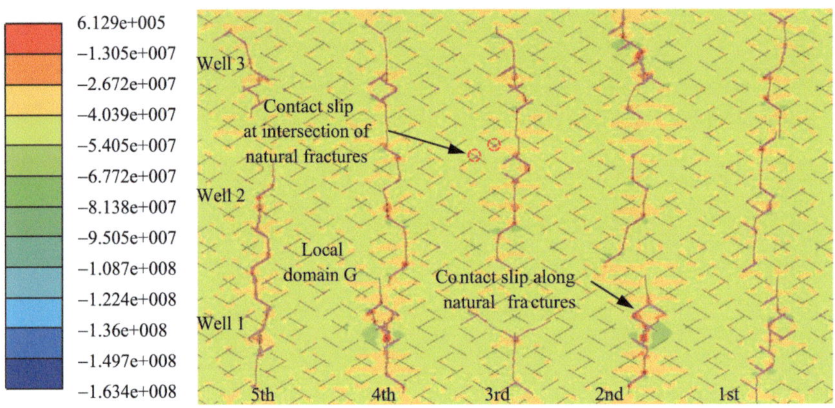

(h) Contact slip events of SC-CO_2 fracturing in naturally fractured reservoir

Fig. 8.12 (continued)

to the computed results shown in Fig. 8.9, the fracture length in SC-CO_2 fracturing is greater than that in slick water fracturing. However, the induced microseismic events of SC-CO_2 fracturing did not significantly increase compared to slick water fracturing. Therefore, SC-CO_2 fracturing can improve the fracturing effect and increase production, but may not simultaneously trigger additional microseismic events.

8.5 Conclusions

This investigation yields the following principal findings:

(1) This study establishes thermal-hydro-mechanical coupled numerical models for multi-well SC-CO_2 fracturing, incorporating natural fractures and multi-well configurations to analyze fracture network connectivity, shear stress shadows, and induced microseismic events. Slickwater and SC-CO_2 fracturing fluids are

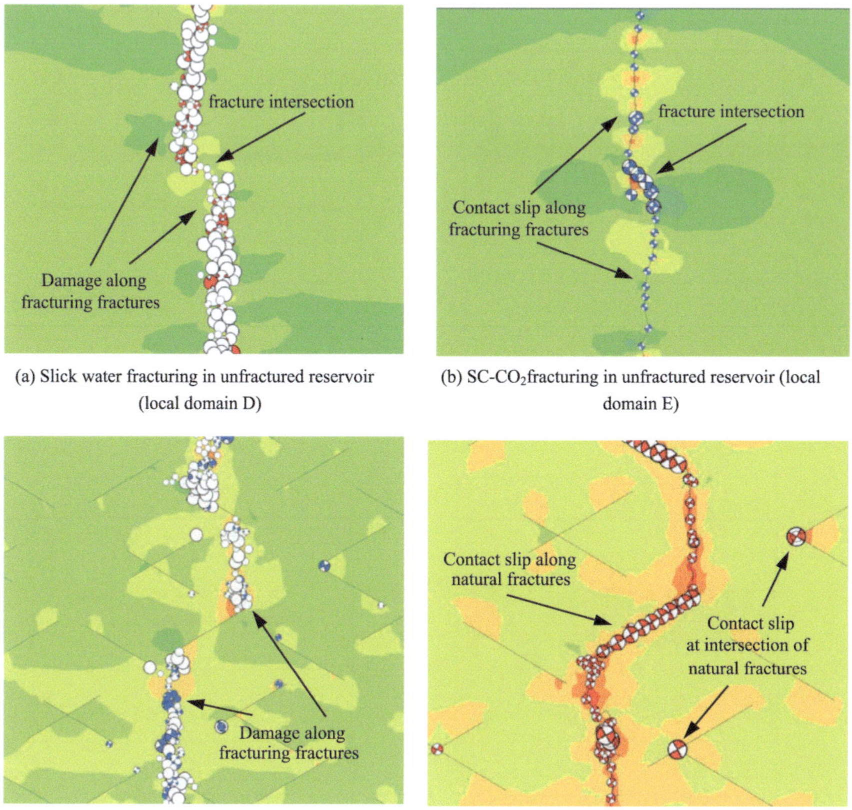

(a) Slick water fracturing in unfractured reservoir (local domain D)

(b) SC-CO_2fracturing in unfractured reservoir (local domain E)

(c) Slick water fracturing in naturally fractured reservoir (local domain F)

(d) SC-CO_2fracturing in naturally fractured reservoir (local domain G)

Fig. 8.13 Distribution of final damage events, contact-slip events and the first principal stress for selected local domains

comparatively evaluated. Quantitative outputs derived from representative cases include: fracture dimensions, stimulated volume, injection rates, pore pressure distributions, and maximum and accumulated magnitudes of microseismic events.

(2) This study comparatively examines multi-well SC-CO_2 fracturing in unfractured versus naturally fractured reservoirs. Results demonstrate that SC-CO_2-induced fractures deflect along pre-existing natural fractures, ultimately establishing interwell connectivity. Quantitative analysis reveals significantly larger fracture networks in naturally fractured reservoirs using SC-CO_2 versus slickwater, owing to CO_2's unique capacity to bridge macroscopic and microscopic fractures, thereby generating complex interconnected fracture systems.

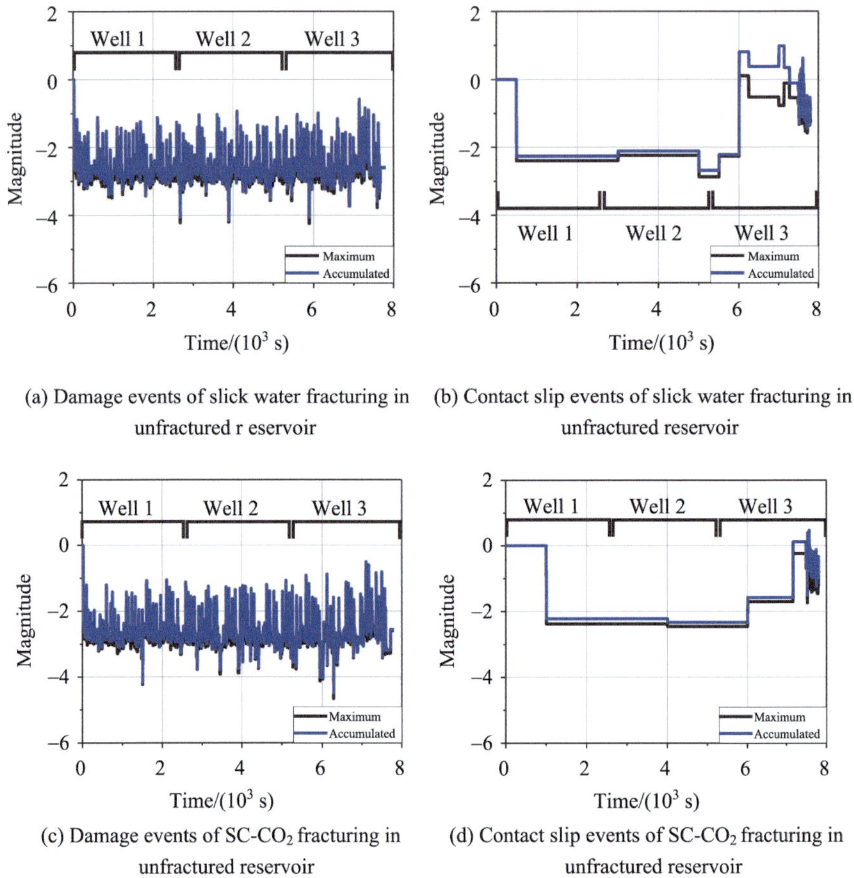

(a) Damage events of slick water fracturing in unfractured r eservoir

(b) Contact slip events of slick water fracturing in unfractured reservoir

(c) Damage events of SC-CO_2 fracturing in unfractured reservoir

(d) Contact slip events of SC-CO_2 fracturing in unfractured reservoir

Fig. 8.14 Evolution of the maximum and accumulated magnitudes of damage and contact slip events

(3) Microseismic damage and contact slip events induced by multi-well fracturing were quantitatively analyzed. In contrast to slickwater fracturing, SC-CO_2 enhances fracture propagation length without proportional increases in microseismic activity; consequently, this method improves stimulation efficiency and hydrocarbon recovery while mitigating the risk of additional induced seismicity.

This study's findings on multi-well SC-CO_2 fracturing establish theoretical foundations for optimizing deep oil/gas extraction designs and mitigating induced microseismic hazards, thereby accelerating field-scale implementation. Notwithstanding these contributions, the underlying mechanisms governing enhanced fracture network complexity remain unexplored; consequently, future research must elucidate the fundamental physics and dominant control factors of fracture network development in SC-CO_2 systems.

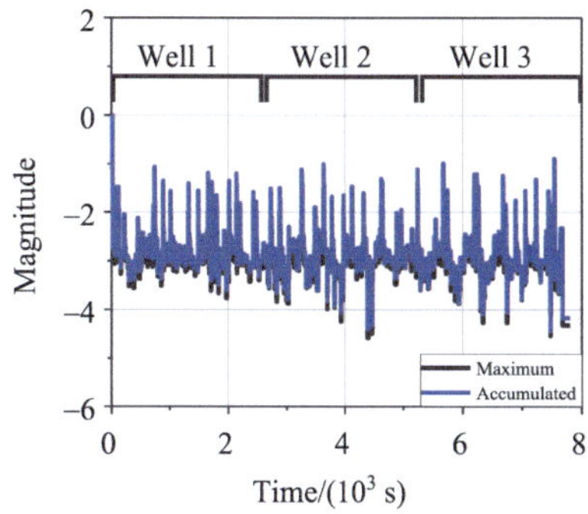

(e) Damage events of slick water fracturing in naturally fractured reservoir

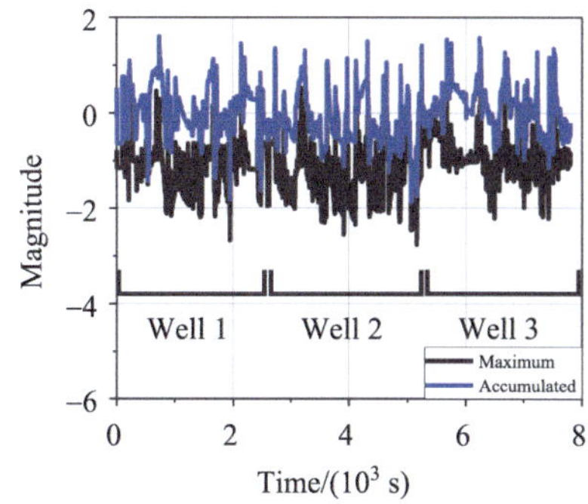

(f) Contact slip events of slick water fracturing in naturally fractured reservoir

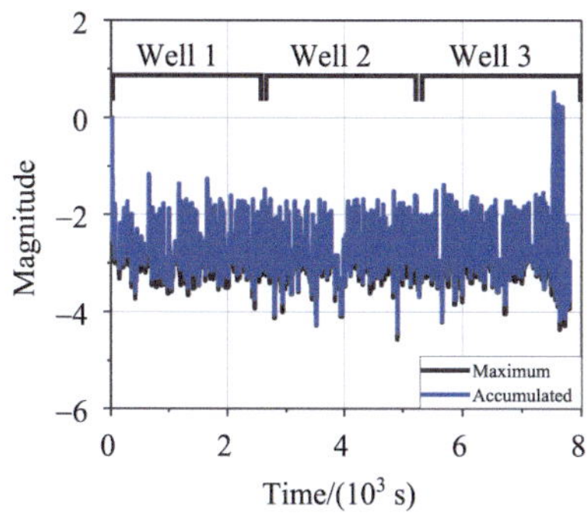

(g) Damage events of SC-CO_2 fracturing in naturally fractured reservoir

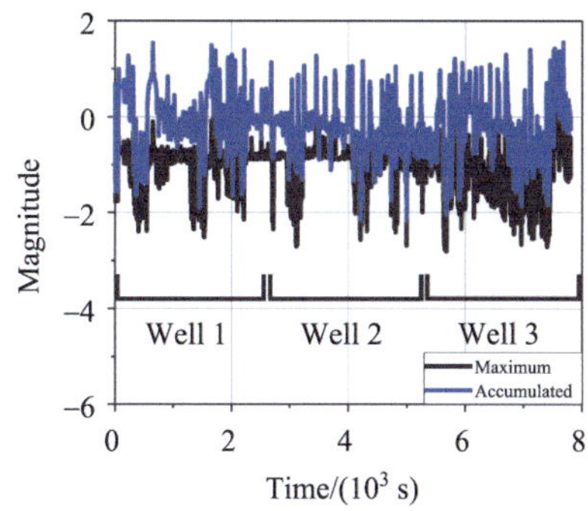

(h) Contact slip events of SC-CO_2 fracturing in naturally fractured reservoir

Fig. 8.14 (continued)

References

Cao, M., Hirose, S., & Sharma, M. M. (2022). Factors controlling the formation of complex fracture network in naturally fractured geothermal reservoirs. *Journal of Petroleum Science and Engineering, 208*, 109642. https://doi.org/10.1016/j.petrol.2021.109642.

Chen, X., Li, Y., Zhao, J., Xu, W., & Fu, D. (2018). Numerical investigation for simultaneous growth of hydraulic fractures in multiple horizontal wells. *Journal of Petroleum Technology, 51*, 44–52. https://doi.org/10.1016/j.jngse.2017.12.014

Dehghan, A., Goshtasbi, K., Ahangari, K., & Jin, Y. (2015). The effect of natural fracture dip and strike on hydraulic fracture propagation. *International Journal of Rock Mechanics & Mining Sciences, 75*, 210–215. https://doi.org/10.1016/j.ijrmms.2015.02.001

Ghaderi, A., Taheri-Shakib, J., & Nik, M. A. S. (2018). The distinct element method (DEM) and the extended finite element method (XFEM) application for analysis of interaction between hydraulic and natural fractures. *Journal of Petroleum Science and Engineering, 171*, 422–430. https://doi.org/10.1016/j.petrol.2018.06.083

Guo, J., Luo, B., Zhu, H., Wang, Y., Lu, Q., & Zhao, X. (2015a). Evaluation of fracability and screening of perforation interval for tight sandstone gas reservoir in western Sichuan Basin.

Journal of Natural Gas Science and Engineering, 25, 77–87. https://doi.org/10.1016/j.jngse.2015.04.026

Guo, J., Zhao, X., Zhu, H., Zhang, X., & Pan, R. (2015b). Numerical simulation of interaction of hydraulic fracture and natural fracture based on the cohesive zone finite element method. *Journal of Natural Gas Science and Engineering, 25*, 180–188. https://doi.org/10.1016/j.jngse.2015.05.008

Guo, T., Wang, Y., Chen, M., Qu, Z., Tang, S., & Wen, D. (2023). Multi-stage and multi-well fracturing and induced stress evaluation: an experiment study. *Geoenergy Science and Engineering, 230*, 212271. https://doi.org/10.1016/j.geoen.2023.212271.

Ha, S., Choo, J., & Yun, T. (2018). Liquid CO_2 fracturing: Effect of fluid permeation on the breakdown pressure and fracturing behavior. *Rock Mechanics and Rock Engineering, 51*(1), 3407–3420. https://doi.org/10.1007/s00603-018-1542-x

Hantschel, T., & Kauerauf, A. I. (2009). *Fundamentals of basin and petroleum systems modeling*. Springer Science & Business Media.

He, J., Zhang, Y., Li, X., & Wan, X. (2019). Experimental investigation on the fractures induced by hydrofracturing using freshwater and supercritical CO_2 in shale under uniaxial stress. *Rock Mechanics and Rock Engineering, 52*(10), 3585–3596. https://doi.org/10.1007/s00603-019-01820-w

Jia, Y., Lu, Y., Elsworth, D., Fang, Y., & Tang, J. (2018). Surface characteristics and permeability enhancement of shale fractures due to water and supercritical carbon dioxide fracturing. *Journal of Petroleum Science and Engineering, 165*, 284–297. https://doi.org/10.1016/j.petrol.2018.02.018

Karimi-Fard, M., & Firoozabadi, A. (2001). Numerical simulation of water injection in 2D fractured media using discrete-fracture model. In *SPE Annual Technical Conference and Exhibition*, SPE-71615-MS. https://doi.org/10.2118/71615-MS.

Li, X., Li, G., Yu, W., Wang, H., Sepehrnoori, K., Chen, Z., Sun, H., & Zhang, S. (2018). Thermal effects of liquid/supercritical carbon dioxide arising from fluid expansion in fracturing. *SPE Journal, 23*(6), 2026–2040. https://doi.org/10.2118/191357-pa

Li, Y., Peng, G., Tang, J., Zhang, J., Zhao, W., Liu, B., & Pan, Y. (2023). Thermo-hydro- mechanical coupling simulation for fracture propagation in CO_2 fracturing based on phase-field model. *Energy, 284*, 128629. https://doi.org/10.1016/j.energy.2023.128629.

Liao, J., Hu, K., Mehmood, F., Xu, B., Teng, Y., Wang, H., Hou, Z., & Xie, Y. (2023). Embedded discrete fracture network method for numerical estimation of long-term performance of CO_2-EGS under THM coupled framework. *Energy, 285*, 128734. https://doi.org/10.1016/j.energy.2023.128734.

Liu, C., Zhao, A., & Wu, H. (2023). Competition growth of biwing hydraulic fractures in naturally fractured reservoirs. *Gas Science and Engineering, 109*, 204873. https://doi.org/10.1016/j.jgsce.2023.204873.

Lu, Z., Jia, Y., Cheng, L., Pan, Z., Xu, L., He, P., Guo, X., & Ouyang, L. (2018). Microseismic monitoring of hydraulic fracture propagation and seismic risks in shale reservoir with a steep dip angle. *Natural Resources Research, 31*(5), 2973–2993. https://doi.org/10.1007/s11053-022-10095-y

Maity, D., & Ciezobka, J. (2019). Using microseismic frequency-magnitude distributions from hydrofracturing as an incremental tool for fracture completion diagnostics. *Journal of Petroleum Science and Engineering, 176*, 1135–1151. https://doi.org/10.1016/j.petrol.2019.01.111

Maxwell, S. C., Waltman, C., Warpinski, N. R., Mayerhofer, M. J., & Boroumand, N. (2009). Imaging seismic deformation induced by hydraulic fracture complexity. *SPE Reservoir Evaluation & Engineering, 12*(1), 48–52. https://doi.org/10.2118/102801-PA

Maxwell, S. C., Chorney, D., & Goodfellow, S. D. (2015). Microseismic geomechanics of hydraulic fracture network: Insights into mechanisms of microseismic sources. *Leading Edge, 34*(8), 904–910. https://doi.org/10.1190/tle34080904.1

Patterson, R., Yu, W., & Wu, K. (2018). Integration of microseismic data, completion data, and production data to characterize fracture geometry in the Permian Basin. *Journal of Natural Gas Science and Engineering, 56*, 62–71. https://doi.org/10.1016/j.jngse.2018.05.025

Peng, P., Ju, Y., Wang, Y., Wang, S., & Gao, F. (2017). Numerical analysis of the effect of natural microcracks on the supercritical CO_2 fracturing crack network of shale rock based on bonded particle models. *International Journal for Numerical and Analytical Methods in Geomechanics, 41*(18), 1992–2013. https://doi.org/10.1002/nag.2712

Ran, Q., Zhou, X., Dong, J., Xu, M., Ren, D., & Li, R. (2023). Study on the fracture propagation in multi-horizontal well hydrofracturing. *Processes, 11*(7), 1995. https://doi.org/10.3390/pr11071995

Ranjith, P., Zhang, C., & Zhang, Z. (2019). Experimental study of fracturing behaviour in ultralow permeability formations: a comparison between CO_2 and water fracturing. *Engineering Fracture Mechanics, 217*, 106541. https://doi.org/10.1016/j.engfracmech.2019.106541.

Rockfield Software Ltd. (2016). ELFEN TGR User and Theory Manual. United Kingdom.

Ru, Z., Hu, J., Madni, A., & An, K. (2020). A study on the optimal conditions for formation of complex fracture network in fractured reservoirs. *Journal of Structural Geology, 135*, 104039. https://doi.org/10.1016/j.jsg.2020.104039.

Seifollahi, S., Dowd, P. A., Xu, C., & Fadakar, A. Y. (2014). A spatial clustering approach for stochastic fracture network modelling. *Rock Mechanics and Rock Engineering, 47*(4), 1225–1235. https://doi.org/10.1007/s00603-013-0456-x

Shen, Y., Hu, Z., Chang, X., & Guo, Y. (2022). Experimental study on the hydraulic fracture propagation in inter-salt shale oil reservoirs. *Energies, 15*(16), 5909. https://doi.org/10.3390/en15165909

Song, Y., Lu, W., He, C., & Bai, E. (2020). Numerical simulation of the influence of natural fractures on hydraulic fracture propagation. *Geofluids, 2020*, 8878548. https://doi.org/10.1155/2020/8878548.

Taleghani, A., Gonzalez, M., & Shojaei, A. (2016). Overview of numerical models for interactions between hydraulic fractures and natural fractures: Challenges and limitations. *Computers and Geotechnics, 71*, 361–368. https://doi.org/10.1016/j.compgeo.2015.09.009

Tan, P., Chen, Z., Fu, S., & Zhao, Q. (2023). Experimental investigation on fracture growth for integrated hydrofracturing in multiple gas bearing formations. *Geoenergy Science and Engineering, 231*, 212316. https://doi.org/10.1016/j.geoen.2023.212316.

Wang, Y., & Liu, N. (2022). Dynamic propagation and shear stress disturbance of multiple hydraulic fractures: Numerical cases study via multi-well hydrofracturing model with varying adjacent spacings. *Energies, 15*(13), 4621. https://doi.org/10.3390/en15134621

Wang, Y., & Zhang, X. (2022). Dual bilinear cohesive zone model-based fluid-driven propagation of multiscale tensile and shear fractures in tight reservoir. *Engineering Computations, 39*(10), 3416–3441. https://doi.org/10.1108/EC-01-2022-0013

Wang, H., Li, G., & Shen, Z. (2012). A feasibility analysis on shale gas exploitation with supercritical carbon dioxide. *Energy Sources Part a: Recovery Utilization and Environmental Effects, 34*(15), 1426–1435. https://doi.org/10.1080/15567036.2010.529570

Wang, Y., Ju, Y., Chen, J., & Song, J. (2019). Adaptive finite element-discrete element analysis for the multistage supercritical CO_2 fracturing of horizontal wells in tight reservoirs considering pre-existing fractures and thermal-hydro-mechanical coupling. *Journal of Natural Gas Science and Engineering, 61*, 251–269. https://doi.org/10.1016/j.jngse.2018.11.022

Wang, Y., Duan, Y., Liu, X., Huang, J., & Hao, N. (2021). Dynamic propagation and intersection of hydraulic fractures and pre-existing natural fractures involving the sensitivity factors: Orientation, spacing, length, and persistence. *Energy and Fuels, 35*(19), 15728–15741. https://doi.org/10.1021/acs.energyfuels.1c02896

Wang, Y., Wang, J., & Li, L. (2022). Dynamic propagation behaviors of hydraulic fracture networks considering hydro-mechanical coupling effects in tight oil and gas reservoirs: A multi-thread parallel computation method. *Computers and Geotechnics, 152*, Article 105016. https://doi.org/10.1016/j.compgeo.2022.105016

Wang, Y., Li, L., & Ju, Y. (2023). Stratal movement and microseismic events induced by multi-well hydrofracturing under varying well spacings and initiation sequences. *Engineering Computations, 40*(7/8), 1921–1946. https://doi.org/10.1108/EC-01-2023-0013

Warpinski, N., Mayerhofer, M., Agarwal, K., & Du, J. (2013). Hydraulic-fracture geomechanics and microseismic-source mechanisms. *SPE Journal, 18*(4), 766–780. https://doi.org/10.2118/158935-PA

Xiong, D., & Ma, X. (2022). Influence of natural fractures on hydraulic fracture propagation behavior. *Engineering Fracture Mechanics, 276*, 108932. https://doi.org/10.1016/J.ENGFRACMECH.2022.108932.

Yang, P., Zhang, S., Zou, Y., Li, J., Ma, X., Tian, G., & Wang, J. (2023a). Fracture propagation, proppant transport and parameter optimization of multi-well pad fracturing treatment. *Petroleum Exploration and Development, 50*(5), 1225–1235. https://doi.org/10.1016/S1876-3804(23)60461-6

Yang, Y., Li, X., Li, X., & Zhang, D. (2023b). Mechanisms of hydraulic fracture propagation and multi-well effect during multiple vertical well fracturing in layered reservoirs with low permeability. *Energy and Fuels*. https://doi.org/10.1021/acs.energyfuels.3c01408

Yao, J., Zeng, Q., Huang, Z., Sun, H., & Zhang, L. (2017). Numerical modeling of simultaneous hydrofracturing in the mode of multi-well pads. *Science China-Technological Sciences, 60*(2), 232–242. https://doi.org/10.1007/s11431-016-0377-y

Yuan, C., & Zhang, J. (2018). A feasibility study of imaging hydraulic fractures with anisotropic reverse time migration. *Journal of Applied Geophysics, 155*, 199–207. https://doi.org/10.1016/j.jappgeo.2018.05.015

Zhang, J., Li, Y., Pan, Y., Wang, X., Yan, M., Shi, X., Zhou, X., & Li, H. (2021). Experiments and analysis on the influence of multiple closed cemented natural fractures on hydraulic fracture propagation in a tight sandstone reservoir. *Engineering Geology, 281*, 105981. https://doi.org/10.1016/j.enggeo.2020.105981.

Zhang, H., Chen, J., Zhao, Z., & Qiang, J. (2022). Hydraulic fracture network propagation in a naturally fractured shale reservoir based on the 'well factory' model. *Computers and Geotechnics, 153*, 105103. https://doi.org/10.1016/j.compgeo.2022.105103.

Zhao, H., Wu, K., Huang, Z., Xu, Z., Shi, H., & Wang, H. (2021). Numerical model of CO_2 fracturing in naturally fractured reservoirs. *Engineering Fracture Mechanics, 244*, 107548. https://doi.org/10.1016/j.engfraemech.2021.107548.

Chapter 9
Summary and Prospect

9.1 Summary

The chapters of the book can be summarized as follows:

(1) In Chap. 1, the research background and significances of deflection of fracturing fracture network disturbed by discontinuity and multiple fractures in rock are well summarized and analysed. This chapter can provide a state-of-art review of the related research and a comprehensive grasp of the research in this field.
(2) In Chap. 2, the morphology of hydraulic fractures is mainly affected by deflection behaviours of fractures meeting embedded heterogeneous and discontinuous geological structures, such as the granules and natural fractures. Evaluating the intersection and deflection behaviours (penetration, diversion, and arrest) and morphology of hydraulic fractures is a key scientific issue to control and optimize the fracturing effects. It is urgent to quantitatively investigate the intersection and deflection behaviours and quantitative morphology of hydraulic fractures in heterogeneous rock masses and analyze the influences of heterogeneity on fracturing effects. To investigate the dynamic intersection and deflection behaviours of hydraulic fractures meeting granules and natural fractures in tight reservoir rock, the numerical models and cases based on statistical modelling and fractal characterization are proposed. The statistical modelling for tight heterogeneous reservoir rock, including statistical Weibull distribution of multi-materials and establishment process of statistical models with granules and natural fractures are developed; the global procedure for statistical modelling, fracture propagation, and fractal characterization is established. Furthermore, using the combined finite element-discrete element-finite volume method, the dynamic intersection and deflection behaviours of hydraulic fractures meeting granules and natural fractures are investigated and analyzed. The results show that the granules have a more inhibitory effect on hydraulic fractures, affecting their propagation; hydraulic fractures are easily connected to natural fractures,

Y. Wang, *Deflection of Fracturing Fracture Network Disturbed by Discontinuity and Multiple Fractures in Rock*, https://doi.org/10.1007/978-981-95-1569-1_9

forming longer fractures, and promoting their propagation. The fractal dimension representing the complexity of fracture network decreases with the increase of the statistical average granule size, and the small granules are more likely to induce the penetration, diversion, and arrest behaviours of hydraulic fractures at tiny scale; the fractal dimension representing the complexity of fracture network increases with the increase of the statistical average size of natural fractures, and the larger natural fractures are more likely to induce the penetration and diversion behaviours of hydraulic fractures at large scale. From the results of this study, the proposed statistical modelling and fractal characterization analysis methods are effective, which lay the foundation for the subsequent establishment of heterogeneous structures in various reservoir rocks, the simulation implementation of hydraulic fractures, and quantitative analysis of fracture network morphology.

(3) In Chap. 3, to investigate the deflection behaviour and fractal morphology of hydraulic fractures meeting bedding and granules, the numerical models and cases with different geometrical configurations and geomechanical properties are proposed. Based on the combined finite element-discrete element-finite volume method and fractal characterization method, the fracture deflection and quantitative fractal morphology of hydraulic fractures considering the influences of beddings and granules are investigated and analyzed. The fractal dimension for the cases of bedding dip angle $\beta = 45°$ is smaller than that under $\beta = 0°$ and 90°; this is because the hydraulic fracture propagation is influenced by the bedding plane, resulting in deflection and propagation along the bedding plane, which affects the degree and complexity of hydraulic fracture network. The comprehensive enhanced geomechanical properties (Young's modulus, tensile strength, and cohesion) in bedding geomaterials of the bedding planes hinder the propagation of hydraulic fractures, leading to a decrease in the complexity of the hydraulic fracture network and the fractal dimension; in details, the larger the Young's modulus is, the more the number of deflections and branches of fractures is, and the more complex the morphology is; the tensile strength and cohesion increase the strength of the bedding planes, which leads to many penetrations of fractures and reduces the complexity and fractal dimension of fracture network. The smaller the granule size of different granule configurations under the same geomechanical properties, the larger the fractal dimension, indicating that the small granule size increases the probability of fracture deflection and complexity of the fracture network. As the geomechanical properties (Young's modulus, tensile strength, and cohesion) of the granules increase to improve their stiffness and strength, it can be seen that the fractures are prone to be influenced by the granules, resulting in many deflections and branches and larger fractal dimension.

(4) In Chap. 4, based on the well-developed dual bilinear cohesive zone model and combined finite element-discrete element method, the dynamic propagation of tensile and shear fractures induced by impact load in rock is investigated. Some key technologies, such as the governing partial differential equations, fracture criteria, numerical discretization, and detection and separation, are introduced

to form the global algorithm and procedure. By comparing with the tensile and shear fractures induced by impact load in rock disc in typical experiments, the effectiveness and reliability of the proposed method are well verified. The dynamic propagation of tensile and shear fractures in the laboratory- and engineering-scale rock disc and rock strata are derived. The influence of mesh sensitivity, impact load velocities, and load positions are investigated. The larger load velocities may induce larger fracture width and entire failure. when the impact load is applied near the left support constraint boundary, concentrated shear fractures appear around the loading region, as well as induced shear fracture band, which may induce local instability. The proposed method shows good applicability in studying the propagation of tensile and shear fractures under impact loads.

(5) In Chap. 5, using the discrete fracture network model, the numerical analysis for center- and edge-type intersections of hydraulic fracture network under varying crossed natural fractures and fluid injection rate are implemented. By varying the level of sensitivity factors, the combined finite element-discrete element method is used, and some typical cases are established to investigate the effects of above sensitivity factor (orientation, spacing, length, and persistence of pre-existing crossed natural fractures, fluid injection rate) on the hydraulic fracture propagation. There are center- and edge-type intersections of fracture network morphologies under varying crossed natural fractures and fluid injection rate. The hydraulic fracture can intersect with the edge of the natural fracture and lead to edge-type propagation, which is conducive for the fracture propagating towards the area farther away from the perforation; in edge-type propagation, when the approach angle between hydraulic fractures and natural fractures is small enough, the hydraulic fractures will be reoriented and activate the natural fractures. The center-type propagation is the result of the intersection of hydraulic fractures and crossed clusters of natural fractures, and The hydraulic fracture may intersect with the natural fracture cluster to form a center-type propagation. Compared with large-scale natural fractures, the small-scale and aggregated center- and edge-type intersections of fracture network morphologies are formed in this study; small-scale natural fractures are more sensitive to the propagation behaviour and final propagation morphology of hydraulic fractures, and are more sensitive to the change of fluid injection rate. The length of fractures during the fracturing process is positively correlated with gas production, and the fitting curve is derived to quantitatively characterize the relationship. For the sensitivity factors (orientation, spacing, length, and persistence) of natural fractures and fluid injection rate, the formed center-type intersections of hydraulic fracture network may generate long fracture length, which is prone to improving gas production; when the hydrofracturing scheme is designed, it is crucial to actively promote the center-type intersections of hydraulic fracture network based on the morphology of natural fractures. When small-scale natural fractures form small-scale and aggregated center- and edge-type intersections of fracture network, the increased fractures gather together to form the clustered

low pressure area and will not continue to increase gas production; the small-scale and aggregated fractures that may play a redundant or even negative role in improving gas production are formed.

(6) In Chap. 6, the combined finite element-discrete element method was used to simulate multi-well hydrofracturing under different distributions of perforation clusters. To study the multi-well hydraulic fracture propagation under the distribution of crossed perforation clusters, the case of parallel distribution of multi-well perforation clusters was set for comparison, and the well-correlated and connected long hydraulic fractures were analysed. The results show that the multi-well crossed perforation cluster distribution can effectively avoid the connection between long hydraulic fracture and the fracture of adjacent wells, such that a single fracture can propagate efficiently, and a complex fracture network can be formed. The variation trends of the fracture propagation and stress shadows under different well spacings and fracturing sequences of crossed perforation clusters were similar to those under parallel perforation clusters. Under a multi-well crossed perforation cluster distribution, the stress disturbance caused by the fractures of the first well reduced, and the degree of fracture deflection also reduced.

(7) In Chap. 7, the numerical models for analyzing the deflection of fracture networks and gas production in multi-well hydrofracturing utilizing parallel and crossed perforation clusters are established. The combined finite element-discrete element method is used, and the coupling effects of solid, fluid, and temperature fields are considered. The numerical cases of multi-well hydrofracturing under varying fracturing scenarios and well spacings are designed and simulated. For deflection of fracture networks in multi-well hydrofracturing utilizing parallel perforation clusters, due to the alternate fracturing scheme, the enlargement of relative well spacing weakened the disturbance of stress and the deflection of fractures; as the well spacing increases, the fracture deflection also weakens, and all the fractures propagate almost straightly; once the crossed perforation clusters scheme is utilized, the number of connected fractures in each horizontal well is reduced, and the deflection of each fracture has also been greatly reduced. For gas production in multi-well hydrofracturing utilizing parallel perforation clusters, the deflected fractures in simultaneous fracturing avoid crossing, which leads to the occurrence of fractures in more zones of the reservoir and is beneficial for the migration and production of tight unconventional gas in the reservoir; as the spacing between wells increases, the disturbance of fractures stimulated by each well decreases, and the stimulated volume of reservoir by fractures increases, which is conducive to the recovery and production of tight unconventional gas; compared to the parallel perforation clusters scheme, the crossed perforation clusters scheme can provide more gas production and larger fracture length.

(8) In Chap. 8, the numerical models considering the thermal-hydro-mechanical coupling effect in multi-well supercritical CO_2 (SC-CO_2) fracturing were established, and the typical cases considering naturally fracture and multi-wells were proposed to investigate the intersections and connections of fracturing fracture

network, shear stress shadows, and induced microseismic events. The quantitative results from the typical cases, such as fracture length, volume, fluid rate, pore pressure, and the maximum and accumulated magnitudes of induced microseismic events, were derived. In naturally fractured reservoirs, SC-CO_2 fracturing fractures will deflect and propagate along the natural fractures, eventually intersect and connect with fractures from other wells. The quantitative results indicate that SC-CO_2 fracturing in naturally fractured reservoirs produces larger fractures than the slick water as fracturing fluid, due to the ability of SC-CO_2 to connect macroscopic and microscopic fractures. Compared with slick water fracturing, SC-CO_2 fracturing can increase the length of fracturing fractures, but it will not increase microseismic events. Therefore, SC-CO_2 fracturing can improve fracturing efficiency and increase productivity, but it may not simultaneously lead to additional microseismic events.

(9) In this chapter, all chapters in the book are summarized, and the prospect for future work is introduced.

9.2 Prospect

Based on the research work presented in this book, further works should be implemented:

(1) **Deflection of fracturing fracture network, stress disturbance and control under the influences of discontinuity in rock.** The discontinuity properties of reservoir rock, such as bedding, granules, and natural fractures, affect the deflection of fracturing fracture network, in which the disturbance of *in-situ* stress field is the key influence factor to analyze the mechanisms of the fracture propagation. In the future, the engineering implementation may control and optimize morphology of the fracturing fracture network by evaluating the disturbance of the *in-situ* stress field in extraction of unconventional oil and gas reservoirs.

(2) **Deflection of fracturing fracture network, induced microseismic events, and oil and gas production evaluation under the influences of multiple fractures formed in multi-well multistage fracturing.** In multi-well multistage fracturing, the fracturing fracture network is affected by the interaction of multiple adjacent fractures, in which the interaction behaviours may induce fracture deflection, microseismic events, and even affect the recovery rate of oil and gas production. In future research, it is crucial to investigate the inherent mechanisms between the deflection of fracturing fracture network, induced microseismic events, and oil and gas production evaluation under the influences of multiple fractures.

(3) **Numerical technologies for multiscale fracture propagation, deflection, and geological storage in SC-CO_2 fracturing.** SC-CO_2 fracturing is a potential unconventional oil and gas extraction technology, which can conveniently form

multiscale complex fracture networks and is beneficial for the flow and exploitation of reservoir oil and gas; it is essential to clarify some key issues, such as the multiscale fracture propagation and deflection behaviours, variations of fluid temperature and pore pressure, and CO_2 phase changes. Simultaneously, SC-CO_2 fracturing may implement geological storage, which has practical significance for achieving the goal of reducing carbon emissions. The above SC-CO_2 fracturing process involves challenging issues, such as multiphysical fields coupling, multiphase changes, and multiscale fracture propagation, which require the development of appropriate numerical methods, models, and technologies.

MIX
Papier aus verantwortungsvollen Quellen
Paper from responsible sources
FSC® C105338

If you have any concerns about our products,
you can contact us on
ProductSafety@springernature.com

In case Publisher is established outside the EU,
the EU authorized representative is:
Springer Nature Customer Service Center GmbH
Europaplatz 3, 69115 Heidelberg, Germany

Printed by Libri Plureos GmbH
in Hamburg, Germany